P.S.&W.Co
No1 EX

JAS

HAND HEWN

JACK A. SOBON

MASTER CRAFTSMAN AND ARCHITECT

HAND HEWN

THE TRADITIONS, TOOLS, AND ENDURING BEAUTY OF TIMBER FRAMING

Storey Publishing

The mission of Storey Publishing is to serve our customers by publishing practical information that encourages personal independence in harmony with the environment.

EDITED BY Deborah Balmuth, Hannah Fries, and Michal Lumsden

ART DIRECTION BY Carolyn Eckert

BOOK DESIGN AND TEXT PRODUCTION BY Stacy Wakefield Forte

INDEXED BY Christine R. Lindemer, Boston Road Communications

TEXT © 2019 BY JACK A. SOBON

Storey books are available at special discounts when purchased in bulk for premiums and sales promotions as well as for fund-raising or educational use. Special editions or book excerpts can also be created to specification. For details, please call 800-827-8673, or send an email to sales@storey.com.

Storey Publishing
210 MASS MoCA Way
North Adams, MA 01247
storey.com

Printed in China through Asia Pacific Offset
10 9 8 7 6 5 4 3 2 1

Library of Congress Cataloging-in-Publication Data on file

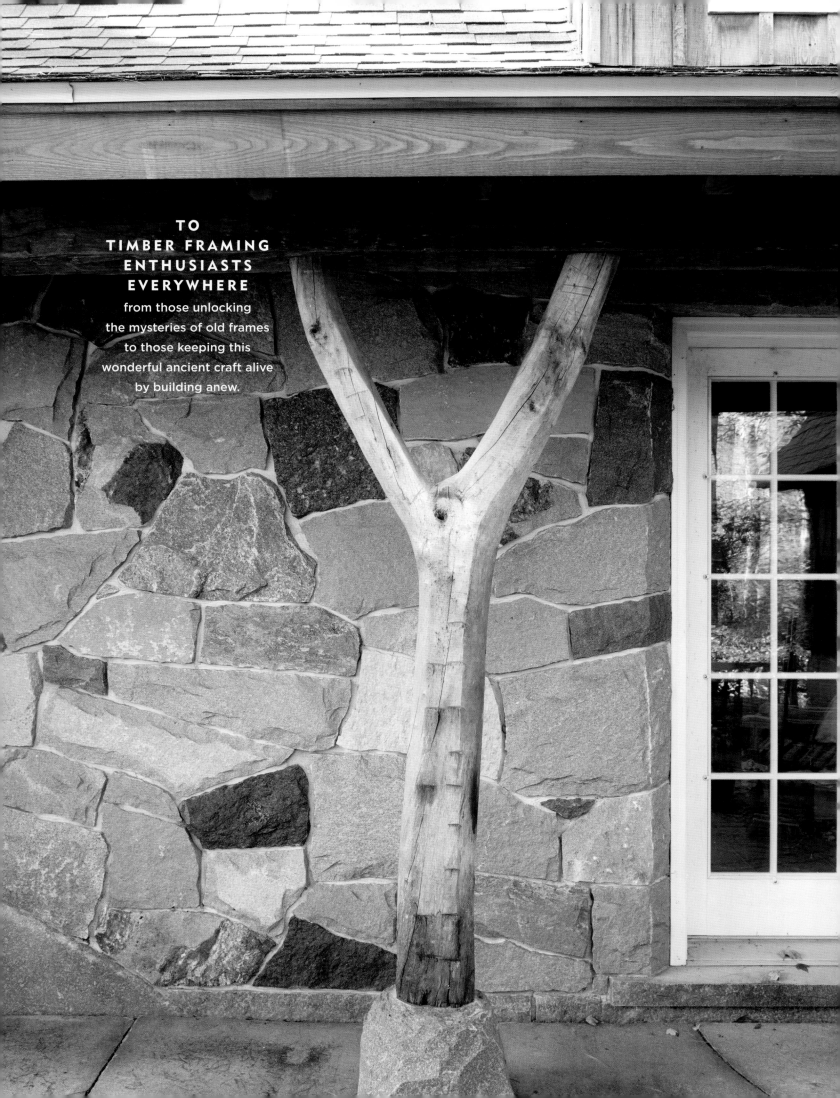

TO
TIMBER FRAMING
ENTHUSIASTS
EVERYWHERE
from those unlocking
the mysteries of old frames
to those keeping this
wonderful ancient craft alive
by building anew.

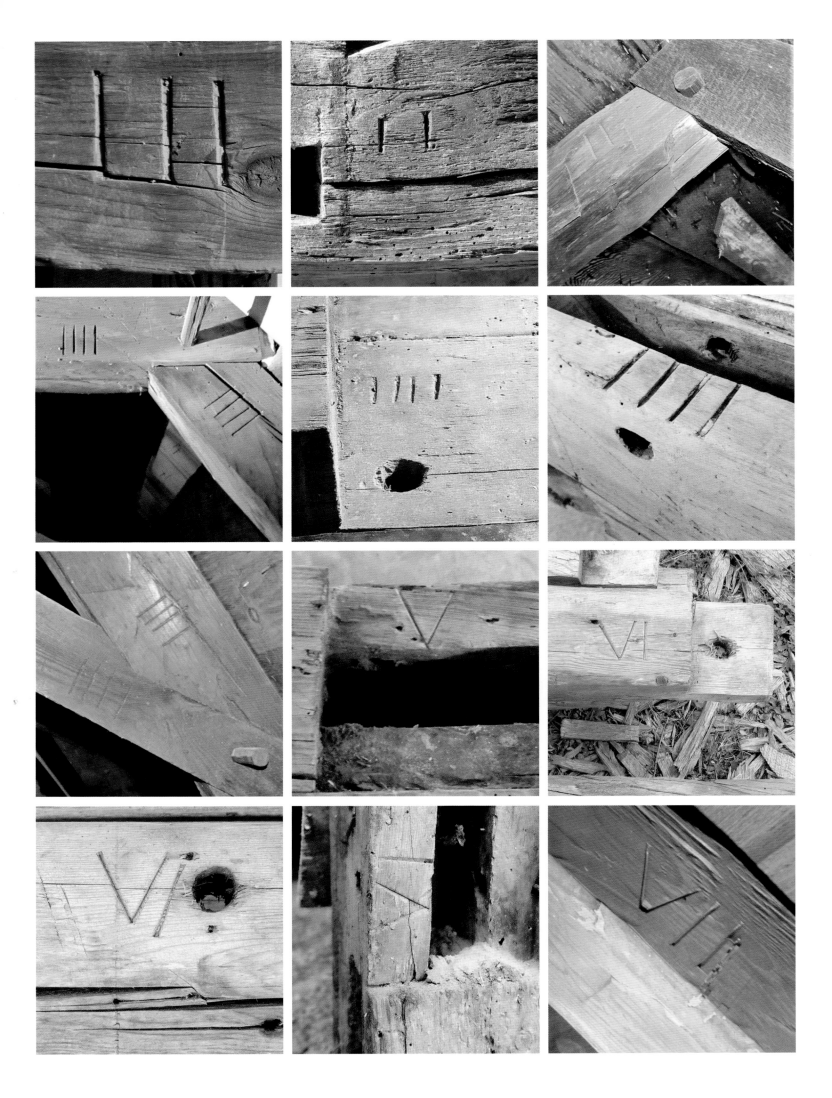

CONTENTS

We Shape Our Buildings . . . Then They Shape Us

I built this modest 16 × 24 timber-framed cottage with Dave Carlon in 1984. It was my first "cruck" frame and features an all-wood interior. For more on crucks, see page 70.

HUMANS CAN LIVE OUT THEIR LIVES in a variety of spaces, adapt to their particular environment, and learn to survive under incredible circumstances. We have been doing it around the globe for eons. But our living environment shapes our lives and determines their outcomes more than we realize. As living beings we respond to light, air, sound, color, smell, and touch in our surroundings. This combination of qualities so influences how we feel that it can elevate our mood and inspire us to be creative and productive and to reach out into the world, or it can sadden or depress us and cause us to withdraw back into our little shell. We often don't see the connection between our mood and our environment and can go about our lives living in uninspiring surroundings, never reaching our potential.

Unfortunately, most of our built environment was designed and constructed by persons not aware of these influences. While building codes are there to protect us from shoddy building practices or unsanitary conditions, a building constructed to the code's requirements still can be quite unsatisfactory for promoting a good life. Even some of the more successful buildings standing today work well only by happenstance. A house plan that works fairly well on one side of the street may be terrible on the opposite side of the street, since turning a building 180 degrees changes the daylighting and exposure dramatically. It is no wonder that when a property changes hands, the new owner may choose to demolish the existing building and replace it with a spiffier version. However, the new structure will very likely have its own shortcomings, and eventually the cycle will repeat itself. Putting some better thought into buildings would save a tremendous amount of wasted resources and effort.

For most of us, our home is the biggest financial investment of our lives. If we are building from scratch, we should not skimp on design; it is too important to our well-being. Or, if we are buying an existing home, we must be vigilant, for there are far too many poorly configured structures out there. It would be a shame to get stuck paying a lengthy mortgage on something that poses, in effect, a liability to your health and happiness. If you watch the real estate listings over time, you will see that certain homes are almost perpetually for sale. There are obviously some problems with them. The owners may not actually blame the house, but they intuitively want to move on. I have found that a good house encourages you to stay put, to invest your life in it, and to put it above other interests. For instance, if the company that employs you decides to relocate its operations, a good house may keep you rooted rather than have you packing to follow the job.

A cozy window seat, framed by crotched tree posts and peeled pole rafters, provides a place to bask in the afternoon sun.

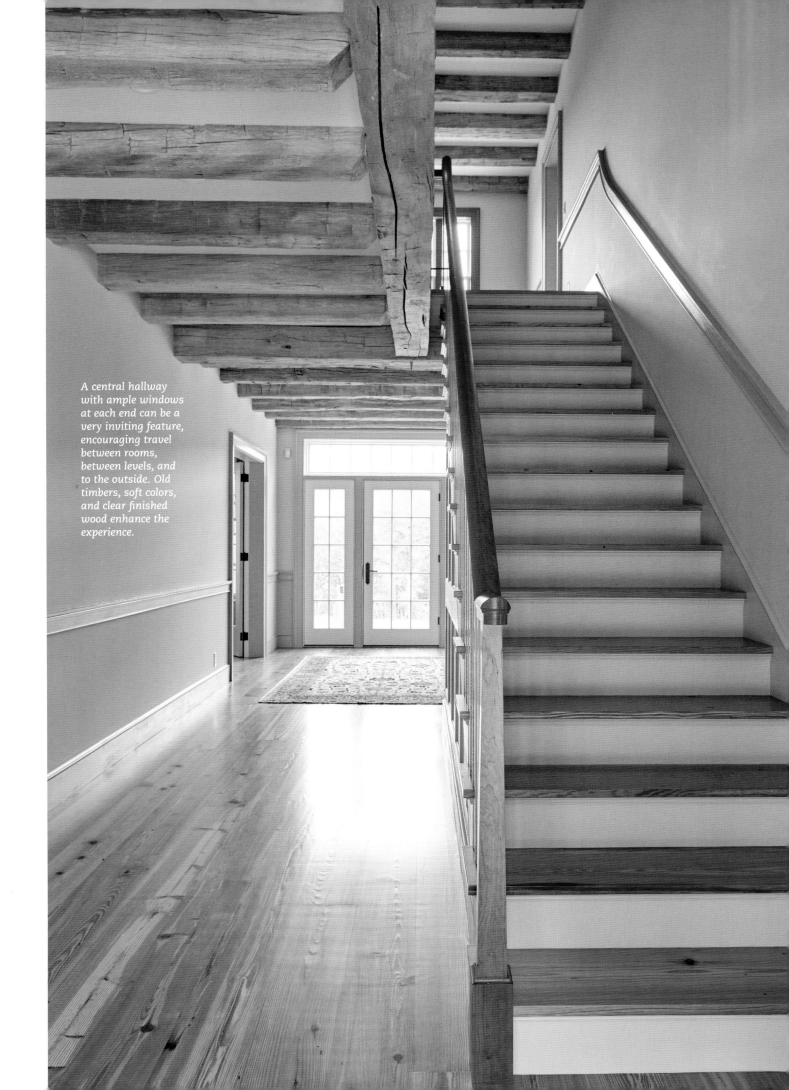

A central hallway with ample windows at each end can be a very inviting feature, encouraging travel between rooms, between levels, and to the outside. Old timbers, soft colors, and clear finished wood enhance the experience.

LEFT *An organic timber frame set against wood walls and ceilings, along with a fireplace constructed of on-site fieldstone, creates a very comfortable space in this guesthouse.*

ABOVE *In this view of the guesthouse loft, the upper parts of the crucks can be seen supporting the roof purlins. The handrail features organically shaped balusters made from small trees of different species.*

Architects are trained to understand how elements like form, light, space, and color affect us. A good architect will create a space that is cozy, nurturing, and relaxing, if it is for a home; healthy and productive, if it is a work space; or perhaps humbling and awe inspiring if it is a spiritual building. While it should certainly meet all the required building and zoning codes, it must also enhance quality of life. I am not advocating for using architects necessarily, but rather for good, thoughtful design that enhances our lives and improves our built environment. If you are not qualified or talented in building design, then involve someone who is. I have lived in a number of different dwellings over my life and, in retrospect, I can see how the various places affected my day-to-day existence. In one place, I literally had to be pried out of bed in the morning, while in another I was the first one up and quickly readied for the day. The only difference was the orientation of the bedroom! In the former example the bedroom was on the northwest corner, where the sun pours in late in the day; the bedroom in the latter was on the east side and received the first rays of the morning. The sun — and the house — were altering my personality. We need to be aware of these influences in our designs.

The Allure of
Timber Framing

Timber-framed rooms are both welcoming and sheltering. Corner posts, a beamed ceiling, and the diagonal bracing are comforting elements. It is best to have daylight entering from windows on two walls rather than only one.

WHEN BUILDING A HOME, we have an almost endless variety of materials to choose from. The options presented in magazines, on television, and at building centers are all choices that some big industry will profit from. If it is advertised, then there is a vested interest behind it. Even the building codes are written to reflect the vested interests of product manufacturers. It seems that only substantial corporations have the necessary capital to invest in the testing and code certification processes that are required to gain approval for new products in construction. Hence, the industry tends to favor the high-tech, manufactured product over the locally sourced and minimally processed one. Of course, the price of the industrial item is higher to cover the costs of promotion.

No other material on earth has been used in as many different ways as wood. From the walls, furniture, and shelving to the firewood, window frames, and timber framing, this scene is an example of wood's unmatched versatility.

In the Catskill Mountains of New York State, the meeting room of the Mountain Top Arboretum education center features a timber frame and flooring of wood harvested from the surrounding forest.

WHAT IS TIMBER FRAMING?

Timber framing is a traditional building system that uses a skeletal framework of both large and small wooden members fastened together with wooden joinery, primarily mortise-and-tenon connections secured with wooden pins. While some connections, especially smaller, nonstructural ones, may be secured with nails, bolts, or other hardware, the majority of connections rely upon wooden joinery. Simply stacking beams on top of posts and fastening them with metal hardware is not timber framing but rather "post-and-beam" construction.

Just as the food in our diets should be whole, locally grown, and without additives, so too should be the materials that surround us in our homes.

The craft of timber framing tends to use locally sourced timber and local hands to create the mortise-and-tenon joints that define it. Though there is a small industry of commercial timber framing, you are unlikely to find timber frame components for sale at your local building supply house. There won't be a shrink-wrapped package of 2-inch tenons on the shelf or a box of step-lapped rafter seats. Nor will you see many ads for timber framing in main-stream publications. Because it is a craft, timber framing doesn't lend itself to the typical methods of the building industry. And if it hadn't already seen several thousand years of use, it probably wouldn't be accepted at all by the code authorities or building industry of today. The fact that there are millions of old timber-framed buildings out there means that they have been "grand-fathered" in.

Though the building industry does very little promotion of timber framing, timber framing promotes itself. Anyone who has lived in or visited a timber-framed house, whether that house is 400 years old or brand-new, becomes a salesperson for the craft. Though not everyone is smitten with the timber framing bug, those who are become lifetime devotees and promoters. I have never seen such enthusiasm anywhere in the conventional building industry.

WHAT IS IT ABOUT TIMBER FRAMING THAT INSPIRES SUCH DEVOTION?

Certainly the allure of the past is part of it. The countless old timber-framed buildings lining the twisting streets of European towns and cities; the great medieval halls of kings and castles with gracefully arching, oaken frames; the temples and pagodas of Asia; and the utilitarian frames of windmills, watermills, and barns — all are at once enticing and inspiring. Standing in and among these buildings, we become part of the past — we are emperors and knights, lords and ladies, yeomen and serfs. When we build in this time-honored tradition, we keep our connection to the past alive.

The craftsmanship inherent in this form of construction is also deeply appealing. Much like a sculptor, furniture maker, cabinetmaker, or other artisan, the timber framer must account for the irregularities of the material, using each piece to its best advantage. This process is neither haphazard nor routine. While machines can be utilized to carry out the more rudimentary procedures, most of the craft relies upon careful handwork with mallet, chisel, and plane. As with other artists and craftspeople, the timber framer must have a vision of the completed work and how each piece being crafted fits into that whole. Usually, components cannot be pre-fitted. The builder must be confident that the pieces will fit when the raising day comes and must hold off on this confirmation until the end, which can be daunting. When it does come, however, after several weeks or months, there is incredible gratification and a high that lasts for days.

TOP *Timber-framed houses feature prominently in the historic sections of many towns in Britain. This street in Lavenham has colorful buildings built with close vertical studding, a symbol of prosperity in its time.*

BOTTOM *Door of the Guildhall, a National Trust property in the village of Lavenham, Suffolk, England*

Then there is the wood itself. Wood is like no other material. Having once been alive, it has qualities and characteristics unmatchable by any factory-produced composites. Wood offers a nearly infinite variety of grain and subtle variations of color, depending on the species of tree, the growing conditions, and the method of converting it to usable material. As with people, no two trees are alike. And within each tree, no two pieces of wood are the same. Wood can be warm to the touch. It is supple yet resilient. It is a good insulator and a good shock absorber, and some species can last a thousand years exposed to the weather. It is the only once-living material on earth suitable for virtually all parts of the house and abundant enough to use on that scale.

Wood is also a renewable resource. On my own 60-acre timber stand, over the last 25 years I have harvested the wood for nearly 50 timber frames of various sizes as well as hundreds of cords of firewood, and there is still more timber — and timber of better quality — standing in that forest now than when I purchased it. When I cut down trees, they are replaced naturally. I never have to replant. Nature (primarily red squirrels) does it for me. What other builder's supply yard restocks all by itself, provides habitat for songbirds, and is a joy to stroll through besides?

Let us not forget the obvious draw of the beauty of the frame itself. It gives the dweller a feeling of confidence when one can see the thick posts supporting the upper floors, the diagonal braces that give the building rigidity in the wind, and the structure supporting the roof above. The assembly isn't hidden or tricking the eye (humans don't like to be fooled). Timber framing is honest construction, and one feels secure within its walls and under its roofs. Though timber frames can be soaring, arching, magnificent, awe-inspiring forms, even the smallest, simplest structure can be graceful. Then

there is the grain and color of the wood, its changing appearance with natural light and artificial light, and the way it changes as it ages. One can stare at wood surfaces and, like peering at the clouds, make out familiar shapes in the grain and knots. Manmade, smooth, monochromatic materials pale by comparison.

Finally, there is the longevity of the timber frame. While it is still susceptible to the effects of decay, fire, earthquake, and vandalism, it will likely fare better than other types of building materials. Countless centuries-old timber-frame structures around the world testify to their resilience. And if an old timber frame building is no longer used and slated for demolition, the frame can instead be disassembled and re-erected on a new site, or the individual parts can be recycled and become a feature in new construction.

Framed in the late 1280s, the Great Hall at Stokesay Castle, Shropshire, England, is the epitome of medieval halls.

Though the detail has lost much of its crispness, this beautiful carving on a sixteenth-century building in Lavenham, England, testifies to the resilience and versatility of wood.

TWO OF THE BUILDINGS THAT I FRAMED earlier in my career have since been dismantled, moved, and re-erected on a new site. Many older timber-framed structures are already in their third or even fourth life! They are ever adaptable to new uses. Working for a contractor involved in the recycling of 200-year-old barn frames is what first got me interested in the craft of timber framing.

So here we have a building system that combines human artistry with what may be the earth's most wonderful material. By timber framing, we are continuing a tradition that goes back thousands of years.

Classic European Timber Frames

Old Town Hall, Esslingen Am Neckar, Baden-Wurttemberg, Germany

RIGHT, TOP Traditional medieval timber-frame architecture of Le Relais Saint Jean Hotel in Troyes in the Champagne-Ardenne region of France

RIGHT, BOTTOM Schiltach, the Bavarian Alps, Germany

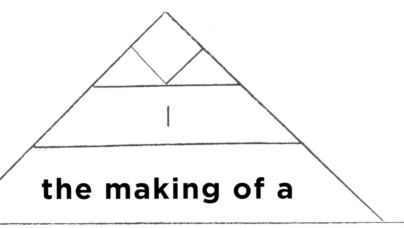

the making of a

TIMBER FRAMER

GROWING UP IN A SMALL TOWN in Massachusetts in the 1960s, it was easy to get swept away in the Modern movement. Everyone was looking forward to the great future ahead: flying automobiles, space travel, and, most importantly, plastics. We were an opportunistic and futuristic society; no one was looking back. In our haste to try every new fad or gadget, we were losing sight of many of the old ways that had served us well for centuries. No one thought twice about abandoning old stuff. I'm sure the preponderance of young people played a big role in that attitude. The number of students in my high school graduating class of 1973 was the school's largest ever, the peak of the post–World War II baby boom.

From Destruction to Reconstruction

AS AN ADOLESCENT, I was fascinated by all the new construction of highways, bridges, and buildings. Preceding much of this new construction was the *destruction* of the old. I spent many hours watching old stone, brick, and timber buildings being destroyed to make way for new shopping plazas, drive-in banks, carwashes, and fast-food chains. The most common way to destroy the old was with a crane wielding a wrecking ball or a clamshell bucket. With the sounds of tearing, splitting, crushing, and crashing down, and the resultant dust clouds, it was quite a spectacle — and watching was one of my favorite pastimes.

My father operated such a crane and was responsible for much destruction. A couple times I sat in his lap while he maneuvered that clamshell to tear open roofs, and a couple times he brought that lovely 1958 American Hoist & Derrick Co. truck crane home for me to play on. I learned quite a bit about how old buildings were built by seeing them torn asunder. It is too bad there was so much destruction, but it was the times. Old was bad, new was good! I guess as a society, we have to suffer some loss before we realize that what is left is valuable. Oh, and there is some wonderful irony here: My dad tore down old timber-framed buildings with a crane when I was a child; I have spent my adult life repairing old timber-framed buildings and constructing new ones and raising them *without* a crane. Because he passed when I was a boy of 10, he wasn't around to witness my lifelong restitution for his acts of destruction.

excerpt from an article in

THE NORTH ADAMS TRANSCRIPT, MAY 1, 1965

EPITAPH FOR A BELOVED AND CHARMING BUILDING

BY GRIER HORNER

JOHN SOBON, the guy who pilots the 60-foot crane for David McNab Deans, swung into the cockpit, his cigar jammed into the side of his mouth, and started easing the levers back and forth skillfully. There are eight levers and two foot pedals and John worked them the same way he's been working them since 1940. Back on this one, forward on that one and the yellow iron jaws that he says go about 2,600 pounds swung into position. The cables on the orange crane's long neck slapped the steelwork and the jaws came crunching down.

In one gigantic metallic bite the six-apartment house that may be the city's oldest building opened up like a hollowed egg hit by a hammer. Bite by bite the brick and dried wood, the pegged hand-hewn beams, the bark covered logs that carried the roof were wrenched from the building in a cacophony of groaning splintering wood and a cloud of dust.

THE NORTH ADAMS, MASSACHUSETTS, TRANSCRIPT

DEMOLITION BEGINS — Wreckers began this morning on demolition of 133-year-old brick-apartment block on Monument Square owned by First Congregational Church. Thus ended controversy in which some members of church had sought to save structure as historical landmark. Church had decided it was not feasible economically to renovate building.

His forehead creased under his steel hat, John Sobon ran his hands over the levers worn down to the bare steel and polished by use.

The jaws arched sideways and up in a fluid motion. Sobon brought them down smoothly over a second-story partition without touching the partition before the jaws locked on it. It pulled out whole and he dropped it into the truck.

And the walls put up in 1832 came tumbling down.

A Summer Job Becomes a Lifelong Pursuit

This fine old Dutch barn was removed to make way for the expansion of a manufacturing facility in Delmar, New York.

WHEN I LEFT FOR ARCHITECTURE SCHOOL in the city, my intent was to eventually live there and design high-rise buildings. Such were the dreams of a small-town boy, but five years of college life in the city and a fortuitous summer job sent my aspirations in nearly the opposite direction. In the summer of 1976 I worked for a contractor, Richard W. Babcock, who dismantled old unused barns, then repaired rotten timbers, replaced missing parts, and re-erected them and finished them off to become upscale homes. It was adaptive reuse of 200-year-old structures, and while a few other, widely scattered individuals were doing the same thing, Richard was on the forefront of this "barns to homes" concept.

DRAWING TO PRESERVE HISTORY

As part of my documenting process, I measured the buildings and created drawings that would be used whether the barn was simply repaired or relocated or it underwent a change of use. Drawings such as these are becoming increasingly important as historic records as countless barns are lost over time.

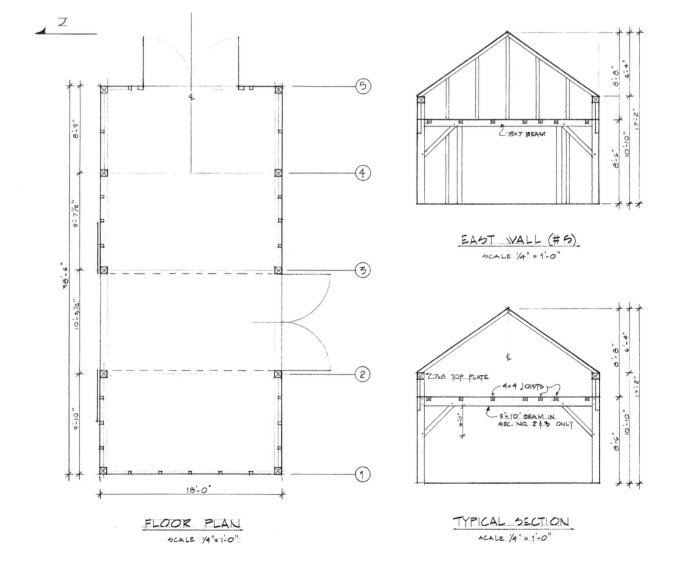

FLOOR PLAN
SCALE 1/4" = 1'-0"

EAST WALL (#5)
SCALE 1/4" = 1'-0"

TYPICAL SECTION
SCALE 1/4" = 1'-0"

It was meant to be just a summer job, but my introduction to this archaic building system — a system that used wooden pins to hold together beams chopped out with an axe alone — became a fixation, a devotion, and a life-long quest. This summer stint ended up changing my architecture direction from high-rise urban structures to wooden, vernacular, and residential ones. It was indeed fortuitous!

In architecture, as in other design professions, school graduates are required to complete three years as an apprentice before they can sit for the four-day licensing exam. I stretched out that period a bit, as I wanted to work for Mr. Babcock as well as a regular architecture firm. I was fortunate to work with him on some very special projects. With our passion for these old barns, we were kindred spirits. We often went "barning" together, searching out the older, more unusual, more magnificent examples of Colonial craft. As we live on the western edge of New England, our barn excursions included much of New England and New York State. I feel quite fortunate to have seen during our travels so many barns, many of which are no longer standing.

Part of my job was documenting the barns by taking photos and measurements, and then drawing the timber frame. I also began designing homes from these barn frames. It is a challenge to adapt a building designed for crops and livestock to function as a home, but when it is done well it is a one-of-a-kind home cherished by its owner.

This barn home was a 1970s in-situ conversion project. An eighteenth-century three-bay English barn had earlier been converted into a New England barn with a basement, a common nineteenth-century improvement. The unusual window arrangement was the result of inserting an arched window over a non-centered barn doorway.

What started as a very small niche of people converting old barns to homes in the late '60s has blossomed into a nearly nationwide trend. As countless barns have succumbed to neglect or have been salvaged and reconfigured as houses, those remaining are ever more precious. Their old hewn frames are in big demand and often are moved hundreds or thousands of miles to be incorporated into rustic dwellings. Even if the building's frame is not reused intact, its parts can be salvaged to be used as repair or replacement parts in other barns, or they can be sawn into boards or planks to finish other buildings.

The old-growth, richly patinated, and honey-colored wood cannot be matched by newly cut, small, second-growth wood. And wood is an inherently recyclable material. Many of the barns we saw had already been relocated once or had timbers in them that were reclaimed from an earlier building. This recycling has been going on for centuries; we didn't invent it. Some of the oldest wooden structures in the world contain timbers from even earlier buildings!

JOINERY: THE HEART OF TIMBER FRAMING

Since examining my first barn, I have been studying and documenting the wooden joinery
that holds old timber-framed buildings together. In the photo below, Dutch anchor beams lie in storage,
awaiting a new life as part of a barn or house. The drawing beneath is of the tying joint in a
1715 English barn in Uxbridge, Massachusetts.

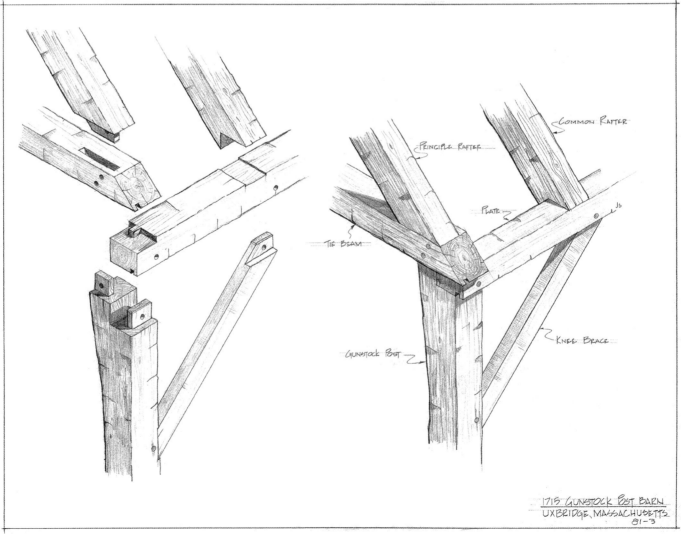

COMMON RAFTER

PRINCIPLE RAFTER

PLATE

TIE BEAM

GUNSTOCK POST

KNEE BRACE

1715 GUNSTOCK POST BARN
UXBRIDGE, MASSACHUSETTS
81-3

THE LEGACY OF RECYCLED MATERIALS

Built in 1131 on a promontory overlooking a branch of the Sognefjord in Norway,
Urnes Stave church features on its north wall elegantly carved components reused from an earlier church
that dated back to 1070. Recycling of materials is not a new concept!

This large, classically detailed home in western Massachusetts makes use of salvaged barn timbers adapted and resized for their new purpose.

BABCOCK BARN HOMES

Below is a piece of promotional artwork for Babcock Barn Homes. Since we worked with actual antique barn frames, re-erecting them as they originally stood, each home became a statement from the past.

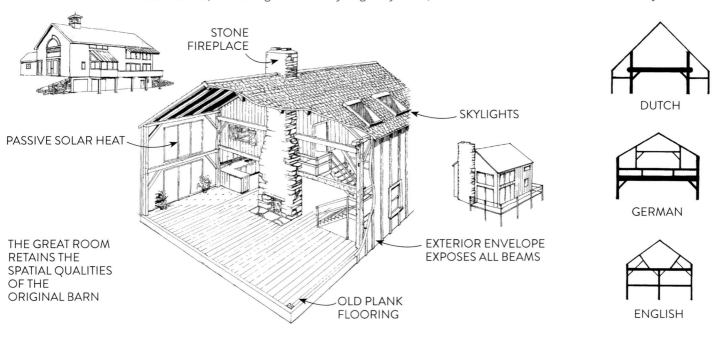

STONE FIREPLACE

SKYLIGHTS

PASSIVE SOLAR HEAT

THE GREAT ROOM RETAINS THE SPATIAL QUALITIES OF THE ORIGINAL BARN

EXTERIOR ENVELOPE EXPOSES ALL BEAMS

OLD PLANK FLOORING

DUTCH

GERMAN

ENGLISH

Mr. Babcock's dream was to relocate period barns from different ethnic areas, creating a sort of living history museum. There would be animal activities, demonstrations, even a timber-framed covered bridge. The polygonal structure in the center is a Dutch "hay barracks," a hay storage system using an adjustable-height roof.

Babcock Barn Museum

Hancock, Massachusetts

The First Cabin

The cabin, with its hand-split pine shingles, provided rustic living arrangements on my woodland property while I worked on constructing a larger dwelling.

THOUGH MY TRAINING was in architecture, my hands longed to build, not just draw. Studying the old timbers, postulating how the surfaces were hewn and the joinery cut all by hand, inspired me to try traditional timber framing. The art of hand hewing particularly interested me. The idea that one could cut down a tree and square it up into a timber using just a simple axe seemed incredible. And yet, I reasoned that it must not have been that difficult or lengthy a process. After all, there were millions, if not billions, of hewn timbers surviving in the United States alone. And what survives today is likely only a small fraction of what formerly existed. Its use surely would not have been so widespread had it not been an efficient process. I needed to know: was it still a useful, relevant method?

I decided to replicate a small, traditional, timber-framed building to better understand how it was done and whether this ancient craft was still useful today. Though I hadn't any property of my own just yet, it was my intention to purchase a few acres of forest to homestead on. It was about this time that I read *Walden* by Henry David Thoreau. In 1845, Thoreau built a little 10 × 15-foot cabin with his own hands on the shore of Walden Pond in Concord, Massachusetts. He writes of his aim in carving out a little homestead and living off the land:

> I went to the woods because I wished to live deliberately, to front only the essential facts of life, and see if I could not learn what it had to teach, and not, when I came to die, discover that I had not lived. I did not wish to live what was not life, living is so dear; nor did I wish to practice resignation, unless it was quite necessary. I wanted to live deep and suck out all the marrow of life, to live so sturdily and Spartan-like as to put to rout all that was not life, to cut a broad swath and shave close, to drive life into a corner, and reduce it to its lowest terms, and, if it proved to be mean, why then to get the whole and genuine meanness of it, and publish its meanness to the world; or if it were sublime, to know it by experience, and be able to give a true account of it in my next excursion.

I needed to build a Thoreau cabin, a little building that I could stay in, on my own land. Since I didn't have any land yet, I also didn't have any trees to work with. I was too anxious to wait; I needed to get started. The logs for my little 10 × 12-foot cabin came from a variety of sources. Some were purchased from a logger in New York State, and some were scrounged during my travels, like a short piece of black walnut I picked up on the side of the road and hewed into a brace. A few pieces came from two Norway spruce trees that I felled in Williamstown, Massachusetts. (See the facing page for The Norway Spruce [A Cautionary Tale].)

I worked up the logs into squared timbers in the backyard of my mother's house. Each log was carried by hand from my pickup, up the front steps, around the house, and up the hill behind the garden. The neighbors undoubtedly thought me an odd fellow, as did my friends who helped me shoulder the heavy logs. I framed the timbers with mortises and tenons right there or, during the colder months, passed them through the basement window and worked them up in my basement shop. All this hand-carrying of logs and timbers was no model of ease or efficiency, nor was it easy on my back, my pickup, or the cellar window! I wouldn't dream of working like that today, but I was young and I was driven, a man on a mission to unlock the secrets of this craft. There was no holding me back.

This little building made use of naturally curving and crotched tree shapes. It was the beginning of a direction and passion that I would later refer to as "Organic Medieval Revival." Two of the posts were crotched, with the smaller fork acting as a diagonal brace. One of the tie beams was naturally arching, and some of the braces were curved. Though straight pieces were readily available and probably faster to frame up, I enjoyed working with, and felt compelled to use, the organic shapes.

I hand-hewed each of the timbers for my 10 × 12 "Thoreau" cabin in my mom's backyard. It was only by using the traditional tools that I would learn the intricacies of the craft.

THE NORWAY SPRUCE

(A CAUTIONARY TALE)

The owner of the trees I felled for my first cabin was a prominent local businessman who wanted more sun to reach his in-ground pool and tennis court area. A double row of Norway spruces between 80 and 90 feet tall stood between the pool and the court. He had already paid a tree service to remove a few of them. Since the trees were delectably straight and the trunks quite free of branches (knots are the bane of woodworkers), I offered to take out two of the remaining trees for free so I could use the wood for timbers.

Arriving on a Saturday morning with a pickup, a chainsaw, and a friend to help, I started the work. Though there was a clear, open space that I could fell the tree into, within five minutes of arriving the first tree was lying across the pool! I hadn't left enough hinge on the stump, the upper branches were entangled with the adjacent trees, and the tree was heavy on one side. The trunk swung around 90 degrees, crushing two chain-link fences and perforating the owner's new energy-saving pool cover. I still remember the rush when that tall tree came crashing down.

After cleaning up the tree and vacuuming the needles from the pool, I headed off with a pickup load of logs. The next day, with tail between my legs, I replaced some chain-link fence components and fixed the pool cover. Later that week, the owner called to see when I would return to get the second tree! After all that, he wanted me to return? Well, he stated, if I didn't go back for the second one, I would certainly lose my nerve.

He was right, of course. He was willing to overlook the events of the previous weekend in order to help me improve my skills and develop my moral character. I returned and felled the second tree without incident, but the legend of the first tree remained fodder for gossip in all the local coffee shops for some time.

From those two spruces, I hewed out a couple of sill beams, three floor joists, a tie beam, a plate, a girt, and a couple of rafters. Oh, I also hewed out a 6 × 12 timber that I later gave as a wedding present to my working partner at the time, Paul Martin, and his wife. They were quite surprised to find it sitting on their living room floor when they returned from the wedding festivities!

I FRAMED THE
TIMBERS WITH
MORTISES AND
TENONS RIGHT
THERE OR,
DURING THE
COLDER MONTHS,
PASSED THEM
THROUGH
THE BASEMENT
WINDOW AND
WORKED THEM UP
IN MY BASEMENT
SHOP.

In the fall of 1980, I purchased an 8-acre parcel in Windsor, Massachusetts, where I set up my cabin. The last few components, mainly rafters, were harvested from trees on the site. In this little building I had timbers of red oak, white ash, sugar maple, Norway spruce, balsam fir, black walnut, black cherry, bigtooth aspen, eastern hop hornbeam, black locust, white pine, American chestnut, white birch, black birch, tamarack, and northern white cedar. There were 20 species in all — remarkable for such a little building.

The foundation was laid of stones garnered from the property and its stone walls. Using a wooden "stone boat," I dragged loads of stone across the pine-needled, duff-covered forest floor. The site was a small clearing that the sun poured into, with a large rock outcropping and a huge, spreading black cherry tree. No trees were felled to clear the area; I just tucked the building into that little opening in the forest. Later, after reading the book *A Pattern Language* (by Christopher Alexander, Sara Ishikawa, and Murray Silverstein), I learned that a building should not be placed directly in the best spot, because that act destroys the beautiful spot. It is better to place it adjacent to that beautiful spot, preferably just north of it, so you don't block the sun from entering it. Fortunately, it was a little building, a tiny clearing, and only a small mistake. I have never forgotten it!

The whole process of building that cabin and subsequently living in it for four years taught me many lessons. I recommend it to anyone contemplating building their homestead or dream house: start small and get to know your land before the clearing and bulldozing begins. Unlike Thoreau, I didn't grow my own food on my land (at least not then), but I did make my living there. While staying in that little cabin, I managed to cut five timber frames, including a frame for my own house that was raised in the fall of 1984.

A HAND-HEWN CABIN IN THE WOODS

Here are the erected frame and the framing plans for my first cabin.

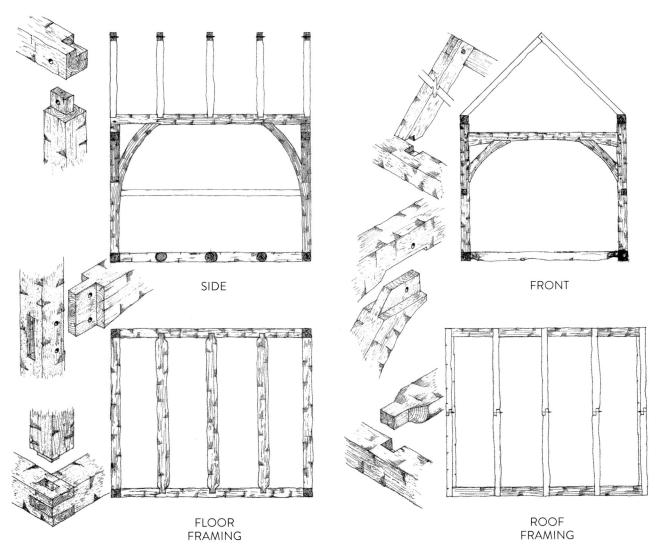

SIDE

FRONT

FLOOR
FRAMING

ROOF
FRAMING

The Beginning of a
Life with Hand Tools

ABOVE AND BELOW This local, family-run mill utilized draft horses for harvesting timber as well as for plowing the farm's fields. Using an old crawler tractor fitted with a boom (a sugar maple log), logs were dragged to the skid trail and then loaded onto a sleigh for transport to the roadside.

THE YEAR 1980 was not only when I purchased my land but also the year that I went into business for myself, albeit part-time. I still worked as a draftsman for an architect, completing my required three-year apprenticeship, so my business began as spare-time and weekend work. My first project was a 12 × 15-foot garden shed with a gabled roof for a friend of the architect. It was framed of all hardwood: beech, sugar maple, yellow birch, and black cherry that I purchased from a unique local sawmill.

Seeing this sawmill for the first time, at the end of a dirt road, perched on a hillside and surrounded by a small farming operation, I felt I was stepping back in time. It was run by two brothers who inherited the setup from their parents and operated it in a low-impact way. Belgian draft horses were used to pull logs out of the forest and also to plow the fields. Some of the equipment was World War II surplus. Though the place looked a bit forlorn, the brothers produced good lumber at a very reasonable cost. In fact, it was the most reasonable lumber available in the area. They were very conscientious about their work. Their logging activities left almost no sign in the forest, since they kept to existing woods roads and trails and used horses, which disturbed the ground very little. Though theirs was not a fast operation, it was a pleasure dealing with the brothers, and I continued purchasing wood from them for many years.

As with my cabin, I decided to use hand tools only on my first paying frame. I had picked up some old tools from flea markets, garage sales, and tag sales. They were typically rusty, chipped, and pitted and had loose, rotten, or outright unusable handles. All had dull cutting edges, of course. I spent many hours in my basement shop sharpening them up and making new handles. As I began using them on the freshly sawn hardwood timber, the tools' binding, pinching, and squeaking told me they still needed more work.

I remember experiencing quite a bit of frustration with each different tool I tried to use. Had I not so much enthusiasm or perseverance, I might have abandoned the whole idea. But with careful observation of how the tools worked — and a lot of fine tuning — things finally started moving smoothly. I got into the rhythm of the hand tools and began to enjoy the work. Even a properly tuned and sharpened tool gets dull after a few hours of cutting and starts performing poorly again. I could see how timber framing and woodworking in general depend on a properly sharpened edge. If I could go back to that time and look at my edge tools, I would probably be appalled, for my sharpening skills have progressed considerably since then. However, they were still sharper than anything I could buy new at the store. Even the finest chisels and plane irons available today require some sharpening and tune-up before they perform at their best.

As the job progressed, I became more efficient and productive. After a slow start, I was ticking off the hours toward the end. My experiment was a success, the work wasn't as hard as I thought it would be, and I was eager to continue in the craft. The client threw a raising party to put up the building, and 50 or so people attended. The little 12 × 15 structure went up quickly and amazed the onlookers. No one could believe I had used only hand tools. Working by hand was unheard of in those days.

Originally I had intended to switch back to power-tool mode after this experiment, but among the attendees at that first raising were my next clients, though I didn't know it at the time. They wanted their own "handcrafted" building. It was a big part of the allure; the process was as important to them as the end product. So, somewhat reluctantly, I agreed to continue the experiment. For this second building — a 24 × 26-foot "horseless carriage shed" (two-car garage) — I partnered with Paul Martin, a fellow employee from Richard Babcock's crew. We handcrafted that frame, laid a stone foundation, raised it with a party, and finished it also by hand. A nice write-up in the local paper's weekly supplement praised our work and the craft of timber framing and extolled the virtues of working by hand. There were now more

TOP *The chisel is the workhorse of the timber-framing craft, and a sharp edge on that chisel is essential to achieving good results quickly. A good craftsman always knows when it is time for sharpening.*

BOTTOM *This 2-inch chisel has seen nearly 40 years of mortise-and-tenon work. The polished patina is from my hands, not from a buffing machine. Below the company stamp are my initials.*

ABOVE, LEFT *The small frame raising was attended by 50 or more excited souls that had never been to a "barn raising" before. Though the building was erected in less than an hour, the merriment lasted until nightfall.*

ABOVE, RIGHT *The 12 × 15 garden shed is eye-pleasing with its geometric proportioning and simple detailing. It sports an oak shake roof and a stone foundation.*

clients for handcrafted buildings. The experiment was over: it was a success. I sold off my power tools. There would be no going back.

I realized that a raising party was the perfect way to advertise and sell yourself; that is, if you don't mess it up. If the raising turns out to be a comedy of errors, it might be your last! This lesson I learned early when attending others' raisings, and so I endeavored to make the structures flawless and orchestrate the raisings to ensure they ran smoothly and efficiently.

Over the years, I upgraded my tools continually. When I came across a better axe, adze, or chisel in my travels, or when I found a type of tool I hadn't tried yet, I bought it. Some had better steel, some fit better in my hands, and some were just more pleasing to the eye. It became an obsession. Because I was young and single, I didn't have to justify my purchases to anyone but myself. These tools were an investment in my trade and would become a part of my life.

Looking back, I have no regrets. A life with hand tools is a good one. My hearing is still acute, and I have all my digits. If I could change anything it would be my lifting techniques back in the early days. Working with big, green, heavy timbers requires using your head as much as your back. Had someone schooled me in the proper techniques early on, and had I not always attempted to move the world, my own frame would be in much better shape today. But that is a difficult lesson to impress upon an invincible-feeling young timber framer.

At many of my raisings, there was so much help that the building seemed to go up effortlessly — and surprisingly fast. Paul and I developed a following of family and friends who were eager to help and attended all our raisings.

Becoming the Sherlock Holmes of Timber Framing

SEVERAL DECADES of using antique tools as they would have been used originally has given me a keen eye when I look at old frames. One might say that I have become the "Sherlock Holmes" of historic timber framing (see chapter 4, "Reading an Old Building," for more on how to interpret the clues). When I began cutting mortises and tenons, I would always compare my finished surfaces with those I saw in old work. Was I using a tool the same way as the original carpenters? If so, the tool marks would prove it.

Another sleuthing question involves the proper sequence for laying out and cutting a mortise. In countless old frames, I find timbers having extra mortises that were cut by mistake. Rather than waste a piece, the builders would use the timber with the mistake. I was pleased that they did! In the pieces that were only partially complete when the error was discovered, one can see a single step in the overall sequence. Eventually, I saw an example of each and every step involved, confirming that I was indeed using the same sequence as the original craftsmen.

Since my introduction to, and education in, the craft has come through working with antique timber-framed buildings, it is only natural that my focus has always been on *resurrecting* the craft, not *reinventing* it. After all, it had developed over thousands of years; undoubtedly, all of the "bugs" had been worked out long ago. "But surely we know better now," you might say. "We have colleges and computers, technology and the internet. We are

One can learn a lot from the mistakes of the past. This timber shows scratched lines for a mortise that was apparently in the wrong place and not cut. The actual mortise is on the vertical face, but the top face shows two V's, which indicate the center of the pin holes.

smarter now!" I have long thought that with my education and experience I might be in a position to make some subtle improvements in timber joinery during my career. For instance, if you examine one of the common joints found in old buildings, you might easily think of a way to improve it structurally. However, for the most part, the variety of joints we find in these old structures are not necessarily perfect in any single respect, but rather moderately good in *all* respects. There are other factors besides strength to consider: There are ease of layout and cutting, ease of assembly, the effects of seasoning, and changes from season to season. Many of these affect the *cost* of the work. Also, changing the configuration of a joint to improve one aspect typically worsens some other aspect. So we must proceed carefully and, when in doubt, revert to what is "tried and true." There is much old work out there, still standing after hundreds of years, to emulate and learn from.

By personally raising over a hundred timber-framed buildings by hand, I have developed insight into how buildings were raised historically. If I cut traditionally designed frames by hand and raise them by hand, over time I am likely to come upon the original and best method that was refined perhaps over a thousand years. Many times in my career I thought I had invented some new technique, only to find evidence of it in a 200-year-old building.

There is a lot to learn in the study of the old. While I have been in a considerable number of old buildings I still find something new, though it might be subtle, in each building I enter. In surveying old buildings, I find so much more to study and record than I did in my earlier years. The more one knows, the more there is to look for. When I travel to other parts of the country or the world, I see so many variations and subtleties, it can be almost overwhelming.

Clues from the Past

Some raising techniques leave evidence of their use that can be read centuries later. This timber has telltale marks of an axe head used to drive the joinery together. The iron poll of the axe easily dented the surface of the pine beam in at least five blows.

When we find mistakes in old work, we gain some insight into the working practices and mind of an eighteenth-century carpenter. For instance, here is a post I once found from a 1760s Dutch house; it had been cut 4 inches too short. The carpenter crafted this tenoned extension to save it.

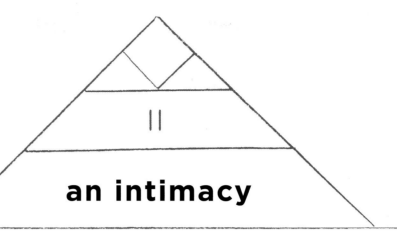

an intimacy

WITH WOOD

MY LIFE'S PURSUIT has been to both understand the craft of traditional wooden building and to bring it forward to use today. I have been working up wooden structures right from the trees since the 1970s, poking around in old buildings whenever I encounter them and spending my leisure time among the trees. I have even heated with wood most of my adult life, and, as anyone who does so will attest, this involves much cutting, splitting, stacking, and general handling. It is safe to say that I have lived a life intimate with wood.

The Building Material Most Like Humans

These 21-inch-wide white pine boards on my own barn were applied in the order in which they were sawed from a single log, a technique called slip-matching (see page 83). Note how the knots line up across the boards.

OF ALL THE COMMON BUILDING MATERIALS available to us now and historically, wood is the most like humans themselves. Having once been alive, wood's characteristics are shaped by its genetics and where it grows. Every tree is different, even when grown on a plantation in straight rows. There are many influences on a living thing that create subtle differences in appearance and workability. As with people, trees also retain scars in their bark and wood that recall their every injury. Much like when I meet an old and revered individual, I am awed when I come upon a venerable old forest monarch that has stood for centuries and exhibits those qualities that only a long, rich life can produce.

This variability tends to confound engineers and building scientists who prefer a building material to have predictable and uniform structural

The Character of Wood

When wood is shaped with sharp hand tools, its grain becomes more visible.

Different species of woods have differing colors and variations in their grain. These hand-hewn 6×6s are, from left: eastern white pine, northern red oak, balsam fir, and (again) northern red oak.

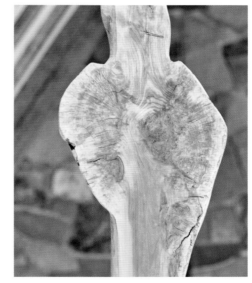

Burls, aberrant growths on trees, contain swirling grain that is prized by woodworkers. Here, a burl has been exposed on a post.

The finest stringed instruments have hand-carved bodies of wood.

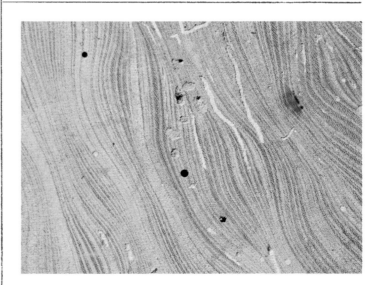

This is a close-up of the end grain of a salvaged elm timber. Note the chevron (zigzag) pattern within the rings that is indicative of elm as well as the holes from powder post beetles and the effect of fluting (muscle-like appearance) of the trunk.

behavior. Unlike the calculable structural values of steel and concrete components, the structural value of wood must be reduced to a fraction of its ultimate strength to account for a variety of strength-reducing characteristics, like knots or sloping grain.

Wood is softer and friendlier than other structural materials; it is not cold to the touch or sharp-edged. When we walk on it, sit on it, or even just brush up against it, it gives a little and is thus easier on our bodies. It is often caressed and admired. Over time, the oils in our skin give it a rich patina. It is the preferred material for tool and implement handles, serving as a transition from the hard steel cutting edge to the soft hand that grasps it. It is even softer acoustically. Sound in a wood-sheathed room resonates more softly and more clearly than in a room encased in concrete, brick, steel, or glass, which can have some disturbing reverberation.

A CHILDHOOD AMONG TREES

My construction career began in the woods, when I was a boy. Using salvaged wood and other building components, we built multi-level tree houses in the woods behind our houses. Nailing directly into the tree trunks was the standard technique (there wasn't much environmental awareness back then), and parents gave us free reign to build our tree forts. They didn't mind us wreaking havoc on the wild landscape as long as we were out of the house. Unfriendly neighborhood scoundrels would destroy our tree houses on a regular basis, so we got plenty of practice rebuilding them. I remember a particularly nice multilevel tree house, perched on top of a hill, that was burned by young vandals. The tree remained as a blackened hulk that forever bore witness to the evil that men (and boys) do.

My favorite forest as a child stood along the perimeter of a rather hilly golf course and contained storybook trees of beech, maple, and hemlock. Once, while playing in an adjacent athletic field during middle school gym class, I remarked to a fellow classmate how the trees on that hillside were the tallest trees in town. Though he scoffed at that suggestion, decades later, I have found none taller than those. In fact, the hillside contains an old-growth forest remnant. Even a child could recognize that there was something special about those particular woods.

The Original Forest

Old-growth coast redwoods (Sequoia sempervirens) in Big Basin Redwoods State Park, California.

WHEN THE EUROPEAN COLONISTS reached these shores, they found forests the likes of which hadn't been encountered in their homeland for many, many generations. Though the Native Americans had been stewards of the land for some time, their modifications to the landscape were gentle. It was a wild land where forests were allowed to grow to their maximum potential. Occasionally Mother Nature would clean the slate with a hurricane or fire, but in many areas, there grew undisturbed forest and some immense trees.

To the early settlers these forests were magnificent, for sure, but they were also a hindrance to their way of life: farming. Each acre of old forest had to be cleared, grubbed, and planted! Though the original primeval or "old-growth" forest of North America is mostly gone, there are a few remaining bits to give us an idea of how it once looked. The rain forests of the Pacific Northwest, the giant sequoias of the Sierras, and the coast redwoods of California are all old-growth superlatives and have among them the tallest, largest, and even some of the oldest trees in the world.

LEFT *The 174-foot "Jake Swamp" tree in Mohawk Trail State Forest in western Massachusetts is currently New England's tallest eastern white pine.*

ABOVE *Old-growth white pine and hemlock in Cook Forest State Park, Pennsylvania.*

Here in the East, and in most of the rest of the forested world, the trees did not reach such grand proportions. There are a few remaining tracts of eastern old-growth forest to help us visualize how it looked to the colonists. The Great Smoky Mountain National Forest, which straddles North Carolina and Tennessee, has 100,000 acres of original forest and is home to the biggest and tallest of many eastern tree species. The Boogerman tree, a white pine that reached 207 feet when measured in 1995, was subsequently shortened by a hurricane to 186 feet. Cook Forest State Park in Pennsylvania has 315 acres of old growth, including the tallest eastern white pine in the Northeast, at 184 feet. Mohawk Trail State Forest in Massachusetts has 612 acres of old growth and the tallest of many New England species, including eastern white pine (174 feet) and white ash (152 feet).

Unfortunately, almost all the eastern old-growth forests that survive do so because of their rugged and inaccessible terrain. The best sites for growing majestic trees, the rich bottomlands, were long ago cleared for agriculture. So, compared to the surviving remnants, the original forest must have been remarkable. There are tales of eastern white pine trees in New England reaching 260 feet high and 10 feet in diameter.

None of the true forest giants of the past survive in old structures today. They were just too big and unwieldy to square up with an axe or fit onto a saw carriage. What we find in old houses and barns are much more modestly sized timbers, planks, and boards. The longest timber that I have verified is a 14 × 17-inch hewn hemlock sill timber, 60 feet long, found in a barn in Middleburg, New York.

Many aisled Dutch barns of New York State had anchor beams (major tying beams, joined to a post) up to 12 × 24 inches and up to 32 feet long. Upon close examination of the anchor beams of a Dutch barn, you can often see that the gable anchor beams are smaller and have larger knots than the interior beams, which are virtually clear (knot-free). It is possible that all the anchor beams (usually four or five in number) came from a single tree. The largest log (butt log) would become a middle anchor beam, while the smaller, more knotty logs from higher up on the tree were used for the gable beams. A 200-foot straight pine could easily yield five 35-foot logs. I suspect it may have been a matter of bragging rights for the owner or builder to declare that all the anchor beams of a barn came from a single tree!

The original forest that greeted the Europeans does survive to a limited extent in the older buildings that survive. If we take a good look at them, we are able to see what the makeup of the local forests looked like and what kinds of wood carpenters preferred to build with.

Old-growth white pine and hemlock trees, Longfellow Trail, Forest Cathedral National Natural Landmark, Cook Forest State Park, Pennsylvania

LARGE TREES, LARGE TIMBERS

This enormous anchor beam in a Dutch barn moved from Fort Plain to Ancramdale, New York, measures 12 × 24 inches in cross section. It is hand-hewn from eastern white pine.

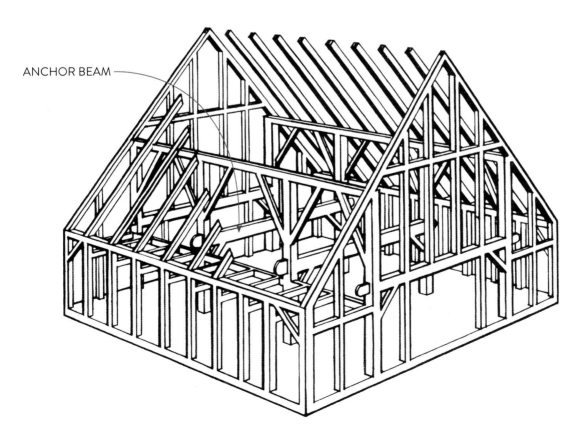

ANCHOR BEAM

FOREST SUCCESSION

The forests of today are mostly *second-growth* forests. The land was cleared of trees and was likely tilled or perhaps just pastured for generations. Or maybe it has always been forested but has been used as a woodlot for centuries. Since it has been cut many times, it has none of the original forest trees still standing. What happens if we leave open land or a forest alone? Will the old growth return?

If you manage a forest for timber production, firewood, wildlife, or just for recreation, it is important to understand forest succession — the changes a forest goes through over time as it progresses from open land to old growth. Before you go into your woods and cut down a single tree, learn to identify the different trees, research your local forest typologies, and begin to understand how they work. You want to make sure that the changes you make by selectively cutting trees will be positive changes.

The following series of illustrations shows how cleared land can become a mature forest once again.

25 YEARS

If open land, pasture, or tilled soil is left alone, wind-borne seeds from nearby trees will settle there. The first trees will likely be pioneer species, shade-intolerant trees that require full sun but will grow quickly. In my area, the pioneer species are white birch, quaking aspen, balsam fir, and white pine. Growing in the open with full sun, the trees will be relatively short as well as branchy and spreading, with trunks that are poorly suited for timber or boards.

50 YEARS

As these trees grow and mature, they provide a canopy of shade. Though most pioneer species are short-lived (white pine is one exception), the canopy they provide allows shade-tolerant species to seed under them. In my area, sugar maple, American beech, yellow birch, and eastern hemlock are examples of shade-tolerant species.

100 YEARS

These slower-growing stems will eventually stretch their way up into the canopy, developing straighter stems with smaller branches and better-quality timber within. The pioneer species will be eliminated as they age because their offspring cannot seed in the shade below; the shade-tolerant species will take over. Each new generation of shade-tolerant trees will grow a little taller, with fewer lower branches and better-quality timber.

200 YEARS

As the makeup of tree species changes, so do the animals and understory plants. After a couple hundred years, the area will have begun to exhibit the characteristics of old-growth forest. If a tornado, hurricane, fire, or other natural disaster or human clear-cut takes place, the process may start all over again.

A Nation
Built on Wood

ABOVE *Logs loaded on railcars in virgin forest, Washington State, between 1935 and 1942.*

RIGHT *Transporting the logs out of the virgin wilderness required building railroad trestles across canyons. This infrastructure consumed vast amounts of timber in itself.*

IT IS NOT AN EXAGGERATION to say that wood built our nation. The proliferation of forests was not only advantageous for building homes and barns but for much, much more. Our forests allowed us to build mighty ships and become a naval power. Timber was used for the ribs, keels, masts, and planking of ships, and live pine trees were tapped to obtain the pitch that made ships watertight. Wood was one of the factors that brought colonists here, and subsequently became one of our biggest exports. Docks, wharfs, and pilings are, of course, made of wood. It was made into barrels and crates to store and ship goods. It was used for wagons, carts, carriages, sleds, and stagecoaches. Early roads through wetter areas were even laid with wood (called corduroy roads). Wood allowed us to fence in vast amounts of land for agriculture and pasture. Most early dams to supply water-powered mills were made of logs and timbers. Oak and hemlock bark were used to tan the hides that provided saddles, collars, harnesses, belts, and whips.

Wood provided the untold number of cross ties that supported the railroad tracks that connected our country, and it was the fuel that locomotives ran on. In 1910, 129 million railroad ties were produced — each one 7 × 9 inches and 9 feet long — accounting for roughly a quarter of our nation's total wood production. The bridges and trestles the train tracks crossed were made of wood too, as were the canal lock gates and the boats and barges on the waterways that traversed below.

As the newer modes of transportation — automobile and aircraft — emerged, they too relied upon wood, which also provided material for shores in mines and scaffolds for construction, as well as a resource for making charcoal. Smelting metal required enormous amounts of charcoal, and the charcoal iron-smelting industry lasted here into the 1940s. A single smelting furnace could require clear-cutting a thousand acres of forest each year, and thus vast areas of forest were cleared just for charcoal.

The vast amount of timber needed for railway ties is hinted at in this photo showing loads of ties being passed down for laying track in Oregon sometime between 1915 and 1920. Each mile of track needed approximately 3,000 ties!

RIGHT A logger jumps across logs in a storage pond at a sawmill.

The forest gave us poles to string our telegraph, electric, and telephone lines. It provided pulp for paper to print newspapers and to fill our schools, libraries, and homes with books. Even cloth fabric (rayon) could be made from wood. Wood gave us turpentine, alcohol, and other extractives for medicine and industry. Maples and birches were tapped for their sweet syrup. Wood provided furniture for homes, schools, and offices, as well as labor-saving devices for home, farm, and industry. It even brought music into our lives with pianos, organs, stringed instruments, and woodwinds. And who could say how much wood was burned in fireplaces and stoves to heat our homes and cook our food over the past four centuries? Where would our country be now if we hadn't had such a resource?

Though we still turn to wood for many of the same uses, it is undoubtedly used less today due to the sourcing of fossil fuels for heating, chemicals, and plastics. We should not forget, however, that wood is a renewable resource available to us wherever forests grow. When fossil fuels are finally exhausted, we will turn to wood once again to help us.

Seeing a Tree from the Outside In

BECAUSE EVERY TREE IS UNIQUE, the wood cut from it also is unique. We have a wide range of tree species on this earth — on my own forestland, I have 26 different species. Each species has its own notable characteristics that vary throughout its range, and indeed within a given woodlot. A tree thriving in rich bottomlands will grow markedly different from one clinging to a windswept mountain ledge. Trees in the open grow quickly and have large spreading branches and extensive root systems. Those in deep forest shade grow slowly, with tall, straight trunks, fewer branches, and fewer roots. Soils, moisture, sun, and wind all affect a tree's growth.

Traditionally, good carpenters worked up their wood right from the living tree. They could tell from looking at the outside of a tree what the wood would be like on the inside. They would walk into a woodlot, eyeing up potential trees for their best use. It was a waste of time and effort to cut down a tree that wasn't usable, so a tree's value had to be assessed beforehand.

The outside of a tree provides clues in the bark, the branching, and the foliage. The diameter of a stem is not always an indication of age, as small, shade-tolerant species can languish in the shadow of their elders for decades, waiting for their break to reach for the sky.

bark

As a tree ages, its bark usually changes. Most trees add a layer of bark each year as they add wood, so a thicker-barked tree of a given species is typically older. Bark with a regular, even, unblemished look means that much of the wood beneath it will be clear of knots or defects. The tree's lower branches probably died off when the tree was still young, or some careful steward trimmed them so they would heal over readily. Decades of growth covered the branch scars, creating the clear wood favored by the craftsperson.

Time and growth can "straighten" a tree by adding more wood in some areas than others: if a straight log is cleft in half lengthwise, often you can see that the tree's center, or pith, is somewhat wavy and irregular.

On the flip side, more irregular bark indicates the branches were larger or have healed over more recently. Damage from storms, falling trees, and animals or humans can leave bark scars too. An experienced eye can tell what the damage is and how far below the bark it resides. It can take decades for a tree to heal over significant damage to roots, trunks, and branches. The wood will decay at the injury, and this decay may spread through other parts of the tree. Much of the trunk's inner volume may be rotten or hollow. By gently

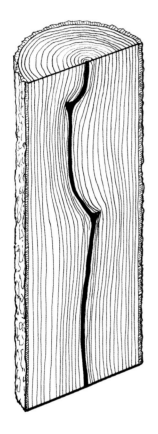

ABOVE *If a straight log is sliced open lengthwise, it is often apparent that the tree started out as a somewhat crooked stem. Each year's successive growth helped to fill out the hollows and "straighten" its profile.*

LEFT *Healthy American beech trees have a smooth, gray bark. This beech is infected with beech bark disease and will eventually die. The bark is first perforated by the wooly beech scale insect and subsequently infected by neonectria fungus, which eventually girdles the tree.*

ABOVE, LEFT *Rotten or hollow trees are best left for wildlife dens and cavity-nesting birds. Birds can help keep forest insect pests in check, so we want to make sure they feel welcome.*

ABOVE, RIGHT *Severe spiral grain is clearly visible in this dead* Pinus sylvestris *tree without bark.*

tapping the trunk, the carpenter can hear the crisp sound of a solid trunk or the dull thud of a rotten one.

If the bark's ridges are not vertical but rather appear to spiral around as they go up, then the wood inside likely spirals too. Spiral-grained wood will twist and warp as it dries. When sawn into straight boards, planks, and timbers, its fibers will not run parallel with the piece. This "cross grain" has much less strength and is to be avoided for most woodwork. A good carpenter will avoid a spiraling tree.

foliage

The amount of foliage on a tree is a good indication of the trunk's proportion of sapwood, the outer few layers of a tree's growth that contains nutrients. A lot of leaves or needles means a fast-growing tree and therefore much sapwood. Sapwood can be detrimental to durable construction, as we shall see later (page 87). Trees from old-growth forests and those in crowded stands will have less sapwood, as they have less foliage. Also, those trees within a given stand that are somewhat suppressed and don't quite reach the canopy will have scant foliage and thus only a modicum of sapwood, and will provide dense, fine-grained, durable timber.

grain

In any tree, the wind blowing against the crown causes bending stresses that are greatest at ground level. To counteract this, the tree's stump grain is particularly sinewy. The wood in taller trees is stronger and heavier not only in the stump but throughout a good proportion of the lower trunk as well. This lower portion typically has smaller-diameter knots and fewer of them. Thus, the lower portion of a tall, branchless trunk is an ideal choice for a carpenter needing a strong beam.

species

The carpenter would also choose different species for different uses. The interlocking grain of elm makes it hard to split for firewood, but that same trait makes it ideal for wagon wheel hubs. Beech, historically a very common wood in the Northeast, is very dense and fine-grained. When two pieces of beech wood are rubbed together, instead of wearing away, the wood develops a polish as though it has been waxed. This makes beech the best candidate for the bottoms of wooden carpenter's planes as well as for gears, axles, shafts, bearings, sled runners, flooring, or any other parts subject to friction and wear. It also can be worked across the grain smoothly and without tearout, so it is superb for turning as well as timber framing. In fact, beech is one of the most common species of timber found in old frames in the Northeast.

BELOW, LEFT *Large logs of beech trees, one of the most common species found in old timber frames of the Northeast*

BELOW, RIGHT *This closeup of a recycled beech post in a new home shows the tiny holes of the powder post beetle. While beech is commonly affected by this pest; it is easily treated with a borate-based solution.*

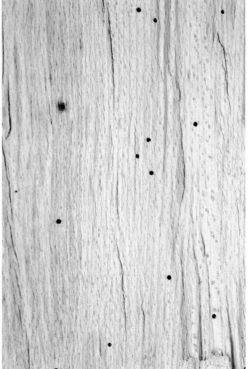

BUILDING GREEN

USING WHAT'S AT HAND

The building industry has a significant effect on the earth's health. The materials we use come from forests and open-pit mines. They are processed using energy, air, and water resources. They are transported great distances, thousands of miles, by diesel-powered equipment, and packaged in plastic and cardboard for convenience. When buildings made with these materials become occupied, they use energy to light, heat, and cool them, as well as more resources to maintain and repair them. If they are constructed on virgin land, they eliminate natural habitat or perhaps good agricultural land. There is no question that buildings are a serious assault on the environment.

Therefore, when we build, the process and the ultimate product should be as benign as it is practical. When we shop at our local building supply, we should choose materials that are certified as "green" whenever possible. This will help the effort. However, the really green materials are not purchased off a shelf

but rather are harvested locally, have no packaging whatsoever, and are minimally shaped or processed by human hands alone.

For those of us living in forested areas, wood is the greenest of all building materials. When we remove a tree, the forest almost immediately begins to replace the lost biomass. If the tree was a hardwood, the stump will often sprout suckers, new leafy stems growing from the existing root structure. These suckers will grow quickly, much more quickly than a tree grows from seed, to fill the gap in the canopy. The ability of trees to resprout after cutting is an ancient form of forest propagation called coppicing. There are forests in Europe that have been managed as coppices for many hundreds of years. While the trees are cut every few years, the root systems are ancient.

In any forest, after a tree is cut down, the neighboring trees will try to fill in the gap with their own new growth. At the same time, with the increase in available sun and water, understory trees

These coppiced red maples started out as sprouts from a cut stump. Because of an established root system, the stems grow quickly, producing decent-quality timber in a relatively short time.

and the tiny little seedlings that were languishing in the shade of the former tree will spring into growth. As anyone who lives in or next to a forest will tell you, it is a constant battle to keep back the forest. If you are a couple weeks behind in mowing your lawn, there are little trees popping up everywhere!

Building with Organic Forms

Organic shapes, with bark removed, form a nice contrast with the straight, squared timbers of this roof structure. While decorative, they are also providing structural support to the principal rafters.

WHEN BUILDING CRAFTERS select the trees from the forest them-selves, they are able to use the parts of the trees that are often not utilized in commercial forest harvesting. Commercial harvesting will take straight trunk sections only and leave the other trunk sections and tops, which tend to be crooked and branchy. For all aspects of timber processing — transporting, sawing, planing, packaging, storing — straight wood is best. However, there are other useful parts of trees that can be quite beautiful and distinctive.

crucks

In one historical building type, the structural core consists of pairs of curved timbers that function as both post and principal rafter. Joined to each other at the peak, they create a rigid triangular cross section. *Cruck* (a derivation of "crook,") framing developed in the 1200s in England as a timber version of the Gothic pointed arch. While it seems to have originated in the upper echelons of society as a formal and decorative system used in great halls, by the 1500s crucks were also used in less fashionable houses and barns. Over

RIGHT *This small, student-built cruck frame features gently curved and tapered blades with arch braces from the cruck blades to the collar. The notches in the upper blades are for purlins.*

BELOW *This small cruck frame in Scotland has jointed crucks utilizing some very organically shaped, nearly round timbers.*

ABOVE *The cruck frame with the widest span is Leigh Court Barn, Hereford and Worcester, UK. It has 11 pairs of cruck blades, with a span between blades of 33 feet, 6 inches. The trees for this barn were felled in the spring of 1344.*

3,000 examples of cruck buildings survive in the United Kingdom, and examples and variants can be found in continental Europe as well.

A cruck truss is composed of two blades (long curved or elbowed timbers) and their connecting members. The cruck blades may run from sill to peak, or may terminate partway down from the peak or partway up the wall. Multiple crucks in a variety of configurations might be seen within the same building. Blades of different shapes required the carpenter to be quite creative in fabricating the cruck's joinery.

The blades might each be squared up from a single trunk, or two blades halved from a single trunk. The builder would start with a curved tree trunk, or with a massive branch at the proper angle. The shape of the blades could be a smooth arc, relatively straight with a curve near one end, an ogee (S-curve) shape, an elbow shape, or an irregular curve. There are even jointed crucks where the blade is made of two nearly straight members joined to create an elbowed form.

While the cruck tradition was largely gone by the end of the eighteenth century, it has seen a revival both here and in the United Kingdom since the 1980s. There is something very appealing about timbers arching up overhead.

RIGHT *Tree crotches are everywhere in the forest, so when we utilize forest resources, it behooves us to use crotches where we can. These crotched trees will be used as posts in a home. The dual-forked example is American beech.*

BELOW *Here I am hand-hewing a black cherry crotched post.*

crotches

LEFT *Crotches figure prominently in this medieval revival home in western Massachusetts.*

Curved, elbowed, and dogleg trunks are quite common in some forest stands. Crotches (forks), while not as common in conifers, are everywhere in decidu-ous forests and can be utilized in timber frames to great effect.

The posts of this house addition are all forked trees that provide not only vertical support to the roof but also diagonal bracing.

If a post with a diagonal wind brace is needed, the carpenter can locate a tree with a large branch in a similar configuration. The natural form will be much stronger and more attractive than a standard brace. Because it grew in that shape, the wood fibers are continuous through the crotch. When the tree was standing in the forest, the crotch sustained thousands of pounds of biomass above it and perhaps more in snow and ice, and it resisted wind loads all the while. We can't match that strength with wooden joinery, steel fasteners, or glue. Nature does it so much better!

round-edge timbers

Round-edge timbers are those that are sawn on one, two, or three sides only. In the days when hewing timbers was common practice, timbers such as first-floor joists were often hewn only on their top side, where a flat surface was necessary to apply the floorboards. The other edges were left in their natural round form. Round-edge components traditionally were used for rafters, joists, and girts, but they can be used for almost any piece. After the timber is sawn on one or more sides, the remaining bark is stripped and the surface smoothed with a tool called a spoke shave, following the natural contour of the curve. If the tree is cut in the springtime as it begins its growing season, the bark will easily separate from the wood and leave a satiny smooth, beautiful surface that any tooling or sanding would only diminish. However, because of the starches and sugars in the cambium layer, the surface should be dried quickly to prevent sap-stain fungi from ruining the appearance of the wood. Round-edge components should be peeled as soon as possible after felling.

BELOW, LEFT (TOP) *The floor joists in my barn are of mixed species, flattened only on their top face to receive the floor planks. For durability, always remove the bark.*

BELOW, LEFT (BOTTOM) *These mixed-species rafters were hewn or sawn on both the tops and bottoms and shaped with squared ends.*

BELOW, RIGHT *Two old joints in round stock: on the left, a joist end, and on the right, a rafter foot with a step-lap joint and tail.*

LEFT *The bark has fallen off this large burl, revealing the swirling grain.*

RIGHT *The back of this handcrafted black cherry chair features a panel cut from a trunk section with a burl. The top and bottom edges are natural.*

BELOW *A closer view reveals the intricate grain lying just under the bark.*

burls

Most woodworkers are familiar with burls, those bulbous growths on trees that can be turned on lathes into highly figured bowls. In commercial forestry, most burls end up rotting in the forest. But they can be used in timber framing as well as in woodworking. When sawed through and planed, burls reveal intricate swirling designs.

book-matching

Even seemingly uninteresting shapes can produce beautiful building components. If you take an average, oddly shaped log and saw it lengthwise down the center of its flattest plane, it will reveal the outline of each year's growth in its grain as well as any branches. When these cut faces are placed side by side, they reveal book-matched grain: one piece is the mirror image of the other. For centuries, cabinetmakers, furniture makers, stringed instrument makers, and timber framers have used book-matching to create handsome, symmetrical designs in wood. In timber frames, book-matching is most commonly used in bracing and crucks.

ABOVE, TOP *Each of these stacks of boards is sourced from a single log; they are piled in the proper order to dry. The "V" saw cut on the ends was made before the squared log was sawn into boards to facilitate putting them back in the proper order for slip-matching.*

ABOVE, LEFT *The siding shown here is slip-matched so all the knots line up across the width of the building.*

RIGHT *A louvered door of curved, slip-matched panels of red oak.*

slip-matching

Slip-matching is a variation of book-matching. Here, a log is sliced into multiple sections of boards, planks, or timbers so that the grain is similar in all pieces. Historically, slip-matching was often used for thinner, plank-like bracing where four, six, or eight matching braces could be fashioned. It was commonly used for flooring and siding. The result is that knots and grain imperfections are no longer distractive elements but rather distinctive features. Instead of chaos, there is symmetry.

Again, utilizing these beautiful, organic elements is nearly impossible when relying on only commercial forestry products when building. To create beautiful, artistic, organic frames, the craftsman must begin the process in the forest.

NOTHING WASTED

When we cut a tree for building or even for firewood, much of its biomass remains behind to replenish the forest. The stump, a little sawdust, the upper unusable portions, branches, and leaves or needles will decompose to help foster new growth. Nothing is wasted. If the logs are squared up in the forest with an axe (hand hewn) or sawn into timbers, boards, or planks with a portable sawmill, then additional biomass is left to decompose or be utilized for firewood.

A 1750s loft or storehouse from
Telemark relocated to the
Norsk Folkemuseum in Oslo, Norway

RIGHT This well-crafted, mortised-and-tenoned
paneled door has probably not seen a coat of
paint for a hundred or more years and yet is
still sound. A modern glued-up, laminated
door would not last a decade without a finish.

Wood in Weather: It Only Improves with Time

SOME OF THE MOST ADMIRED wood surfaces are those that have been weathered by Mother Nature. Over time, exposed wood elements become silvery, with etched grain and rounded edges. The early wood growth (spring growth) is softer than the late wood growth, so it erodes faster, creating minute surface furrows that we treasure. Also, the normal wood grain wears away faster than harder elements, such as knots, leaving them proud of the surface. Radial-cut wood, in which the saw cuts perpendicular to the annular growth rings, weathers better (and looks better weathered) than flat-sawn pieces, which are cut more or less parallel to the growth rings. However, most flat-sawn pieces will have some radial grain nearer their edges.

With exposure to weather and sunlight, wood on vertical surfaces erodes about ¼ inch per century. Vertical, 1-inch-thick, all-heartwood pine siding on a 200-year-old barn will be only about

½ inch thick where exposed, except at knots or where the siding is covered by trim or tucked up under the eaves. If you cut into it, you will see that the wood, though thin, is still sound.

Attempting to seal the wood with paint or varnish is done in vain. Because the wood moves with the building as it racks in the wind and expands and contracts with changes in temperature and humidity, the "sealed" surface will be compromised. Tiny cracks will allow in water that cannot escape. Imagine the sealed wood surface as the roof of a house. If the roof has multiple small holes in it, the first rainstorm will soak the house interior and its furnishings. When the rain stops and the sun comes out, how much of that water will find its way back out those small holes in the roof before the next storm? Probably very little. With the next rain it will only get wetter. The house will stay wet, and decay will set in. Wood lasts much longer if it is allowed to dry out after a soaking. Thus, paints and sealants generally tend to shorten the life of wood.

Penetrating stains function better than sealants because the wood can "breathe." Stains will often make the wood last longer if they contain preservatives. Over time, stain wears away and fades rather than flaking or peeling as paint does, and more stain can be applied without having to scrape the surface. However, when my clients mention that they want a stained finish on their siding, porch, deck, or play set, I have a simple, stark response: However many gallons of stain or preservative they think they will need, they should simply dump it onto the ground, for that is where it eventually ends up. It is far better to design a structure with good detailing and use unfinished wood or wood with natural finishes (such as linseed oil, citrus oil, beeswax, or shellac) than to make up for shortcomings by poisoning the environment.

Wood exposed to the weather (but allowed to dry after it gets wet) will erode at approximately ¼ inch a century and take on a gray, furrowed appearance.

Avoiding Rot, Naturally

WITHIN EVERY TREE there are two kinds of wood — *heartwood* and *sapwood* — and each has different properties relating to durability. In the core of the tree lies the heartwood. In many species, the heartwood is noticeably darker or richer in color, as it contains more extractives, or chemical substances. The only valuable wood of highly prized black walnut and black cherry is the heartwood. Other than providing structural support to the tree and storing moisture, the heartwood is no longer involved with growing or food production. If the tree has any rot-resistant qualities, it is in the heartwood. The sapwood band has no rot resistance, even in species known for this quality, like redwoods and cedars.

The most recent few years of a tree's growth (the outer rings) is the sapwood. While the actual growing portion of the tree is at the interface between the bark and wood, the sapwood is involved with moving the starches and sugars for the tree's growth. In a freshly cut tree, sap will ooze out of this sapwood band, and in pines and spruces, the sticky sap will make itself quite evident.

What part of the tree the wood comes from can make a big difference in how it fares in a building situation. When wood gets wet from rain, the heartwood dries out quickly in the sun, while the sapwood may take days or weeks to dry. Moist wood is a food source for fungi and the larva of wood-boring insects and therefore is a recipe for quick decay. Also, sapwood doesn't hold paint as well as heartwood, making it a poor choice for exterior surfaces of windows and trim. In fact, I would say that sapwood should not be used anywhere on the exterior of a building at all.

The amount of sapwood versus heartwood in a tree is not a fixed ratio. It can vary greatly depending on factors such as species, age, speed of growth, and health. In eastern white pine, the sapwood is roughly the last 15 years of growth. Thus, a 15-year-old pine will be all sapwood. A larger, fast-growing pine with lots of crown foliage might be 50 percent sapwood. In a slow-growing pine with little crown foliage, the sapwood band is narrow: the last 15 years of growth may only have added an inch to the tree's diameter. On old-growth wood specimens, the band is often so narrow that it is insignificant. By the time a piece is squared up for use, the sapwood is effectively gone.

Unfortunately, most lumber used today is from fast-growing trees that provide a quicker return on investment. Their proportion of sapwood is higher and the wood is much less durable. Carpenters often no longer have the wherewithal to identify and select only the heartwood for exterior uses,

At bottom is a cut-off of an eastern white pine log with crystalized sap, indicating the sapwood band. By volume, this log is about 50 percent sapwood. Above, an interior cut shows that the sapwood band also has a fungal stain, commonly referred to as blue stain.

so wood in general has subsequently gotten a bad reputation as an exterior material. Further, the highest-quality boards, those with few or no knots, are typically found in the sapwood band. As a tree grows taller and the lower branches die and fall off, it adds wood over those branch stubs. This newer wood is clear of knots, but because it is mostly sapwood, it will decay quickly. It is no wonder that vinyl, fiber-cement, and composite materials predominate for exterior use today.

What can a responsible carpenter do to remedy the problem?

» Separate out the sapwood lumber to use inside the house for subflooring, flooring, trim work, cabinets, studding, and so on.

» Design exterior wood construction to minimize its exposure to rain, slope surfaces to drain water, and keep siding well above the ground surface. Vertical surfaces will shed water quickly and dry out faster than horizontal surfaces. Higher surfaces will last longer than those near the ground.

» Build eaves with generous overhangs to keep water from splashing against the siding.

» Don't let shrubs, vines, and other plantings grow against wood siding. Areas that get good sun and air circulation will fare better than dark corners that stay damp.

» Use stone instead of wood for features that must contact the ground, such as exterior walkways and stairs. Build stone terraces rather than wood decks. If you must build using wood that will contact the ground or sit out in the rain, use the heartwood of a naturally rot-resistant species, such as black locust, white oak, black walnut, or cedar — and plan on replacing it every 15 or 20 years.

The most common approach today is to ignore good building detailing by using wood that is chemically treated to resist decay. By filling the wood with poison for decay-causing organisms, we increase its durability. Wood railroad ties, utility poles, fence posts, and the like are pressure-treated to be durable. However, there is a price to pay for this kind of durability. We are poisoning our soils and groundwater with these chemicals, blighting the landscape. And what about the health of the people working in these industries? For all involved, it is better to use good detailing and naturally rot-resistant species.

LEFT *These posts are elevated above the patio on granite bases. The wood is black cherry, a rot-resistant species.*

ABOVE, TOP *The porch framing of this 900-year-old Norwegian stave church is elevated on a mortarless stone foundation that allows for air circulation under the timber work and helps prevent rot.*

ABOVE, BOTTOM *Stone can be used for rot-prone locations such as patios and post bases.*

III

history of the craft

IN AMERICA

THOUGH PRECEDED BY a limited amount of timber joinery in the cedar lodge structures of the native peoples of the Pacific Northwest, the timber framing tradition as we know it was brought to North America by primarily English, French, Dutch, and German carpenters who settled along the Eastern Seaboard. At the time of their immigration, timber framing was the standard mode of building with wood throughout much of western Europe. Log construction was favored in the Baltic countries and eastern Europe, where conifers dominated. In areas where trees for building were not so abundant, masonry building predominated. However, the New World, or at least the eastern half of it, was heavily forested and tended to attract those who were used to and skilled in building with wood, both timber framing and log building.

European Influence

LEFT *This reconstructed street using relocated period buildings in the Norse Folk Museum gives us the feeling of how Hammersborg, a suburb of Christiania (Oslo), might have looked in the eighteenth century.*

ABOVE *This storybook setting is the Weald and Downland open-air museum in Singleton, West Sussex, England, where buildings threatened with demolition have been reassembled to create both urban and rural settings.*

FOR A CARPENTER IN EUROPE in the seventeenth or eighteenth century, life and building were constrained by the fact that there was little raw land available and little new building development. Much of a carpenter's work consisted of repairs or additions to existing buildings. Or, for a village carpenter, it might mean repairing fences, gates, and wooden water pumps, or building coffins. New buildings might be predominantly masonry construction with only the roof being timber-framed. Those specialists who worked on larger and more important buildings were required to travel to be able to work on such projects continually. Craft guilds made it possible for masons to travel to different locations, but carpenters' guilds were primarily local. A carpenter traveling to a different area might not be recognized or allowed to work there.

The forests of Europe had been utilized since before recorded history. Little, if any, original forest or old growth remained in most areas. Carpenters and wood users of all types had long grown accustomed to using second-growth material and paying a good price for it. They had to make do with what they had available, and the poorer-quality timber required the skills of

a very good carpenter. The best-quality timber, what there was of it, was likely reserved for the royals and the upper echelon of society.

The New World, on the other hand, was virtually unbounded. There was abundant land for the taking, free high-quality trees, and building opportunities for anyone with a modicum of skill. A carpenter could spend his whole life building entirely new homes and barns! For anyone longing to break free or start anew, this was the place: the New World was a magnet for the young and the restless.

early dwellings of colonial america

The very first dwellings here were temporary shelters or hovels. These crude buildings might have been partially dug, or *bermed,* into the earth and covered with roofs of thatch or bark slabs over sapling rafters. They were heated by open fires on the earthen floor and might have had a clay-daubed wooden chimney to draw out the smoke. These types of buildings were familiar back in Europe. Transient workers such as charcoal burners constructed their own temporary dwellings while they worked the land.

Settlers of eastern European or Baltic origin often fashioned temporary dwellings of horizontal logs. As settlements advanced inland and westward, the new

arrivals could board with others while constructing their first buildings. If they were farmers, their first permanent "framed" building might have been their barn, since sheltering their livestock, their livelihood, was essential.

While these settlers built according to the traditions of their homeland, there were some almost immediate changes in building construction brought on by conditions here. For instance, in England, timber frames were mostly exposed, both inside and out. The spaces between timbers were filled with a woven framework of small twigs or riven laths (wattle) and then plastered with an organic mixture of cow dung, chopped straw, clay, and lime (daub), and finally finished with a rendering of animal hair and lime. Referred to as "wattle and daub," this construction was labor-intensive but utilized readily available free or inexpensive materials. It functioned well and lasted sufficiently long in the mild English climate. In America, wood was abundant and inexpensive. Siding of wood clapboards was much faster to apply and held up better to driving rains and freezing winter weather than a wall of wattle and daub.

roofs and floors for the new world

Roofs in England were commonly made of thatch (reeds). Thatch could be harvested sustainably from marshy areas but required a lot of time and labor to harvest and install. It also required a steep roof pitch to work properly. At "common" pitch, just under 50 degrees from horizontal, rainwater will follow the thatch fibers so that the roof sheds the water. A thatched roof can work for decades if the pitch is steep enough, but it is not a "waterproof" material like slate, tile, or wood shingles. Although reeds grew here in abundance, thatch was utilized less, again because wood was so plentiful. A wood-shingled roof is faster to build and can have a flatter pitch.

Roofs were modified not only by lessening the pitch but also by changing the framing. *Hipped* roofs and *half-hipped* roofs are common in England (see photos at left). Though the carpentry is more complicated with a hipped roof, this was offset by using fewer long, straight rafters, a somewhat precious commodity there. With the inexpensive and high-quality timber available here, the carpenter's time and labor became the precious commodity. While thatching requires the rafters to be spanned by closely spaced laths or battens on which to tie the thatch, wood shingles are better applied to fully boarded roofs. Since boarding and shingling a hipped roof is more work than a simple gabled roof, again expediency won out and gabled roofs predominated.

Another type of roof that became common here due to the preponderance of wood and the need for expediency was the *board-on-board* roof. Instead of common rafter framing, in which closely spaced rafters extend

TOP *This hipped, thatched roof is on a crofter's cottage at the Ryedale Folk Museum, UK.*

BOTTOM *This roof type, where a gabled roof has its ends clipped, is referred to as a half-hipped roof or occasionally as a jerkin head roof.*

Pegged Barn Floors

In the New World, barn threshing floors were thick planks pegged to the floor framing with square pins.

In my own English barn, I have pine floor planks pegged with riven, square, white ash pins.

from the eave to the peak, purlins were added horizontally across widely spaced principal rafters. Over the purlins, boards ran vertically from eave to peak. Another layer of boards was laid over the joints in the first layer, creating a sort of board-and-batten roof. Though it required many long, good-quality boards (no knotholes or sapwood), it was fast! It could be completed in a day or two by two carpenters — and it would last 15 to 20 years, as long as a shingled roof that required many hours of splitting and shaping the shingles, blacksmiths to make nails, and a week or more to lay.

Similarly, floor construction also changed in the New World. In England, the first floor is what we call the second floor here. Their first floor is called the "ground floor," for it was typically on the ground in early buildings and may have been made of packed clay or flat stone slabs. With wood at a premium and the climate relatively mild, an earthen first floor in England was very practical, and it worked for houses as well as barns. In America, with its plentiful wood, elevating the floor above the ground made it drier, warmer, and easier to clean, and it allowed for a root cellar below for storing produce over the winter. While stone floors were used for threshing grain in barns in England, wood-framed barn threshing floors became the norm in the United States.

Scribing: The European Method

THE SYSTEMS OF TIMBER-FRAMING carpentry brought here from Europe varied slightly with their countries of origin, but all relied on setting out and fitting components together in two-dimensional assemblies. Timbers that are hewn out with an axe or sawn by hand or at a mill will typically vary in size within each piece and among pieces; they may not be square or straight along their length. Carpenters had to account for these inconsistencies in all but the crudest assemblies. Their techniques involved a process called *scribing*, or scribe rule, where the irregularities of one mating surface are matched on the end of the member joining it.

I am most familiar with the English method of scribing, so the following description adheres to that approach. Though there is little in the way of written accounts of Dutch and German scribing systems, the surviving framing indicates a style similar to the English. The French system uses a full-size drawing (chalk lines snapped on a floor) on which to overlay timbers. In this French system, timbers are stacked in their respective locations above the snapped lines and, using a plumb line for a vertical reference, the abutments are marked.

In denser urban areas, components were likely precut off-site in the builder's work yard, and then transported to the site and erected. In rural areas, everything was more likely done on-site, reducing the transport costs each way. When scribing timber components, a three-dimensional structure is thought of in terms of its two-dimensional assemblies: floor frames, wall frames, cross frames and roof frames. In America, the first-floor frame, which is basically a full-size floor plan of the building, is typically scribed and assembled first. If cut on-site, it would be set on the foundation, squared, and leveled. The subsequent floors, walls, cross frames, and roof frames would then be positioned and scribed over the floor frame. If this floor frame is properly sized, square, and level, it minimizes additional measuring and squaring of the rest of the frame. It is akin to laying timbers over a full-size drawing (as in the French system).

Each assembly laid over the floor frame requires a two- or three-step scribing process. First, principal components (posts, ties, plates, and so on) are laid on and their joints scribed and cut. Mortises (but not tenons) are pre-drilled for pin holes. The joints are assembled, square and level, as tightly as is practical and secured with temporary iron drift pins, tapered pins with a hooked head that allows for easy withdrawal.

The pin hole location in the mortise is traced on the tenon, and the joint withdrawn enough to bore out the tenon's portion. However, the tenon's pin

Scribing Pin Holes

In scribed buildings, the pin holes in the mortises are pre-bored before assembling the parts. The joint is assembled as tight as is practical, and the hole's location is marked on the corresponding tenon. The tenon is then withdrawn, and the hole is offset and bored. On this tenon (above), just to the right of the hole, is a series of prick marks indicating the mortise hole's location. If there were a gap where the joint came together when the hole was pricked, the carpenter would add that gap amount to the amount of the offset. This accounts for the big offset seen here.

The joint only has to be withdrawn enough to bore out the tenon.

hole is offset a bit closer to the shoulder in order to create a *drawbore* (also called draw pinning, or pull boring). A tapered wooden pin driven through offset holes will draw the joint tightly together and keep it there as the frame is raised, maintaining tightness even as the timbers dry and shrink.

Then, secondary members (braces, girts, studs), which are framed between the primary timbers, are laid on, scribed, and cut. The assembly is taken apart as needed to install the secondary pieces. Then the framing is reassembled, leveled and squared, and again secured with drift pins. Perhaps a tertiary set of timbers (door headers, smaller framing) is then laid on and the process is repeated.

by the numbers

When the assembly is finished, the components are numbered, typically using Roman numerals. The pieces may be numbered from left to right, north to south, west to east, or front to back. Each carpenter would have his preference. The marks would be incised with a chisel, gouge, or a timber scribe (also called a race knife).

Rather than number in sequence from one to several hundred, carpenters used techniques to simplify the numbering. For instance, they might use

THE SCRIBING PROCESS

This 1763 house formerly stood in West Springfield, Massachusetts, and measured 29 feet 7 inches by 39 feet. The following steps show how the frame of the house was scribed, each step being composed of at least two, sometimes three sub-steps. First, primary components are laid out, scribed, cut, and assembled. Then, secondary members are laid on, scribed, cut, and assembled.

1

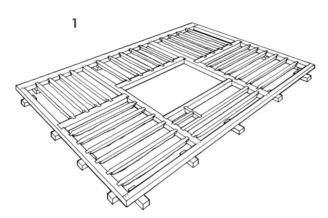

2

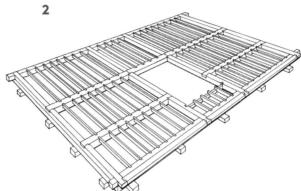

The first step is to scribe, cut, and assemble the first-floor structure, known as the "plan" of the house. It is composed of sills, sill girders, and floor joists. The joists are logs hewn only on their upper face. This structure is set up on blocks in a convenient working area on-site. It is squared up and leveled, since it will serve as a base and template for the scribing of all the other assemblies. The large opening in the center is for the fireplaces, hearths, and central chimney. The smaller opening is for the stairs to the cellar.

The attic floor is scribed next. It is composed of plates, tie beams, summer beams, and joists. The edges of the timbers are lined up with the sills below to ensure its proper size and squareness. Along the front of the house (lower right), the tie beams extend out 8 inches to carry the flying plates supporting the decorative cornice. The central opening is for the chimney mass and stairs to the attic. For clarity, the timbers of the first floor below are not shown.

3

After it is numbered, the attic floor frame is removed. The front wall of the house is assembled next. It includes a plate, tapered posts, braces, and studs. Again, by aligning it with the square floor frame below, the builder makes sure it is scribed square to create a plumb wall when erected. Blocking under the members keeps the top of the framing on a level plane. The larger openings are for the windows and doors. When completed, the front wall frame would be removed and the process repeated for the rear wall, aligning it with the opposite side of the floor frame. For a single-story house (a Cape), the front and back walls can be done simultaneously.

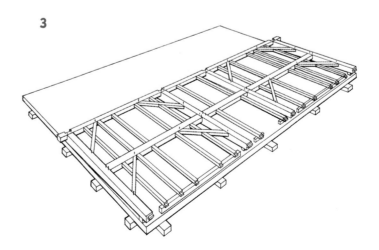

4

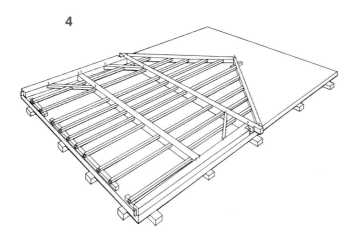

Next, the four cross frames or bents are scribed. Each end wall has posts, ties, braces, studs, and rafters. The posts and the attic tie beam have already been scribed in previous assemblies. When complete, the bent is removed and the process repeated for the remaining three cross frames.

5

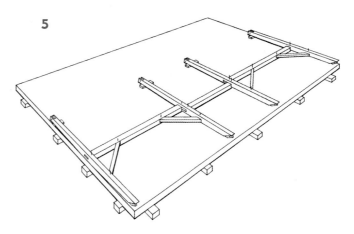

The sloped planes of the roof are next. They include principal rafters, purlins, and braces. In this particular house frame, the smaller common rafters would be cut to a pattern and thus are not part of the scribing setup. The front roof plane, shown here, is done first, then the rear roof plane would follow.

6

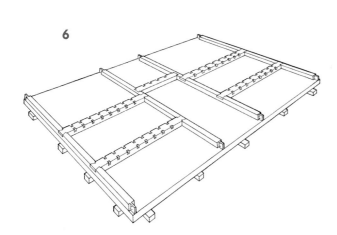

Lastly, the second-floor framing is set up and scribed. It includes the tie beams and summer beams. Since the individual floor joists for this frame were not numbered, they were likely cut to fit after the frame was erected.

7

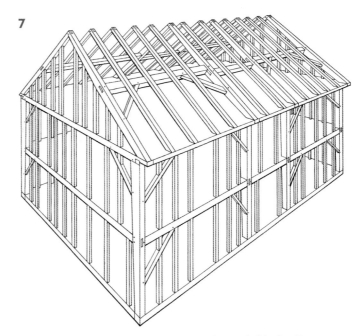

Here is the finished house frame (with the floor frames covered for clarity). The roof system is made up of principal rafters and purlins, with bracing only on the front roof slope.

Matching Up

Simple chisel numbering: For each mark, two cuts at about 45 degrees remove wood to create an incised cleft. The girt on the lower left has its number at the opposite end.

This is a marriage mark scribed with a race knife across the joint. The circle denotes which cross frame it is part of.

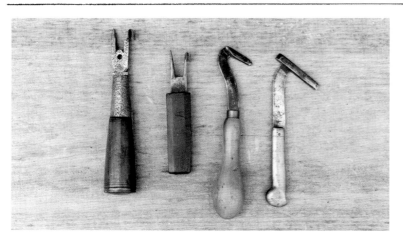

Shown are several timber scribes in my collection: The two on the left, probably eighteenth century, have a compass point to draw arcs as well as a folding cutter for straight marks. The one on the far right has a cutter that folds into the brass handle to protect its edge.

All the wall components of this eighteenth-century Dutch house are numbered with the race knife but have an additional letter — S, N, E, or W — denoting their compass orientation. For some reason, the "S" is backward.

Sometimes numbering is done on the faces of mortises (as seen here) and the sides of tenons and is concealed after assembly. This number four is a series of four marks made with a gouge.

In this eighteenth-century barn, the members are numbered consecutively, using a race knife. This brace is numbered 45. Each crossing is 10, and each meeting is 5.

a 1-inch-wide chisel on the north side and a 2-inch chisel on the south. Or there might be an extra mark or tag (flag) on one side. Using a race knife, a builder might use a half circle on the first cross frame, a whole circle on the next, two half circles on the third, and two whole circles on the fourth. On the longitudinal framing, he would typically use straight numerals, no circles.

The framing was then disassembled and the timbers set aside. Some principal components such as posts, plates, and ties would be brought back for scribing in other assemblies. In each assembly, the most important face of the framing, the *best face*, was up. The best face (also known as *layout face, reference face, upper face, fair face*) was typically the top side of floors and roofs, and the outside face of walls. These faces were required to be flush and even to receive sheathing or flooring boards. Interior cross frames would have a face designated by the carpenter, but in three-bay barns and houses, they typically faced the center bay (the chimney bay or, in barns, the center threshing bay).

the english tying joint

English-American framing, up until about 1800, typically made use of a special joint called the *English tying joint*. Here, the plate, post, tie beam, and principal rafter all join together in a three-dimensional junction utilizing three mortise-and-tenon joints and a half-lapped dovetail joint. In a three-bay barn or house, there are typically eight of these joints.

The post top is typically wider to provide two tenons, one for the plate and one for the tie beam. To create the extra width at the top, the post may be cut with a slight taper from bottom to top, have a gently curved flare, or have a jowled, or *gunstock,* shape. While it may look like this joint design is primarily for structural purposes, it appears that its configuration was mostly to facilitate the scribing process. The joint allows you to pre-assemble each of the three two-dimensional frames that it joins together: the attic floor (or, in a barn, the plan frame at eave level), the longitudinal wall frame, and the cross wall frame (or bent) with its rafters. Each can be assembled and secured with pins or joinery.

Some carpenters continued using the English tying joint in its original form, while others experimented with modifications to improve its efficiency. There are a great many variants of it to be found, including some with notches instead of dovetails, some that have the post turned flat in the longitudinal wall, and some with no laps, only mortise-and-tenon joinery. For expediency, carpenters sometimes utilized lap-jointed members rather than all mortise-and-tenon joints. When scribing lap-jointed pieces, the piece is laid on, scribed, cut, and inserted without requiring any disassembly and reassembly of the framing. This saved considerable time and effort.

TOP *The English tying joint has been variously described as ingenious, confounding, structurally advantageous, and structurally deficient. The post, plate, tie, and principal rafter all join intricately at a single juncture.*

BOTTOM *In my own house frame, I used naturally flared posts sawn on three sides to make eight English tying joints.*

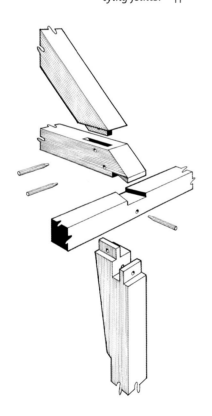

For the scribing of each joint, there are two methods to choose from. The most commonly used method is called *tumbling*, a simple and direct way of transferring length between timbers and angles of abutment.

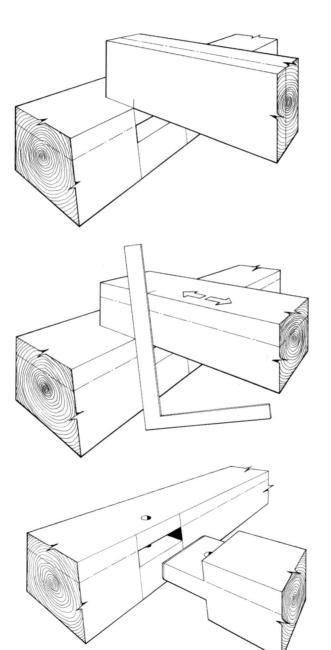

TUMBLING METHOD

Tumbling was the process used to scribe a timber between two other timbers, a very common situation. Timbers are first lined (chalk lines snapped) to indicate the joinery plane. The mortises are made before the timbers are set up. The face of the mortise is dressed to a flat plane. The timber to be scribed is placed above its mortises, turned on edge so that its top face is aligned with the side of each of the mortises. A small mark is made at each end in line with the top edge of the mortise face; this indicates the length between the two timbers at their top surface.

The timber is then rotated (hence the "tumbling") to lie flat in its final position above the mortises. A square, or any straight parallel rule, is held against the mortise face and alongside the timber to be tenoned. The timber is then moved by tapping until the length mark aligns with either side of the square and the line scribed. This is repeated on the opposite side. The whole procedure is then repeated on the opposite end of the timber.

The lines on the sides of the timber are connected across the tops and bottoms, and the tenons cut. The joints are assembled and the pin holes marked for the draw bore. By this process, the piece is cut to the exact length and mated perfectly to the angles of the mortised faces.

DOUBLE CUT METHOD

The other method, used primarily for the top tenons of the gunstock post, is the double cut, or cut-and-try method. With this technique, a tenon is cut shy of its shoulder by a determined amount, then it is assembled into its mortise and scribed to fit with dividers or a compass. Finally, the shoulders are recut to the scribed marks and the joint is reassembled, hence the "double cut."

When tenons are double-cut they can be scribed to follow the irregularities of the mortised member, as seen here. The wane (rounded bark edge) on the mortised member has been fixed to a flat facet.

This eighteenth-century post tenon exhibits both the saw kerf from the initial cut and the chiseling of the shoulder on the second cut. The chisel used had a marred edge, leaving distinct grooves.

The Advent of the Square Rule

In this photo from Tuscola County, Michigan, circa 1865, a master carpenter and his crew of seven apprentices are framing hand-hewn timbers for a barn using the square rule. The workmen are, from the left, posing with a chisel and mallet, an adze, another adze, a saw, and a boring machine (Snell Mfg. Co.); the master is next with an awl in one hand and a framing square in the other; then another apprentice poses with a Snell boring machine, and another with a saw.

AROUND THE TURN OF THE nineteenth century, there was a major change in the timber-framing process. The older system of scribe rule framing brought to America from Europe was replaced by a simpler system not requiring any preassembly of components. This new system, called the *square rule*, created interchangeable parts by reducing timbers to consistent sizes at the joinery. Though hewn and sawn timbers vary in size and may be bowed or twisted, the joinery can be standardized for expediency.

Within every rough timber, which may be bowed, twisted, or varying in size, the carpenter envisioned a somewhat smaller timber that was straight, free of twist, and of exact size. The timber would then be reduced to that smaller size only at the joints where it counted. So a rough 10 × 10-inch timber might have the joints cut to a 9 × 9-inch timber. A square-ruled timber frame is easily identified by the consistent reductions at the joints.

Having cut a number of timber frames by hand, using both scribe and square rules, I have found the square rule to be about half the labor of the scribe rule. The huge amount of time saved by this system caused it to spread quickly. With the exception of some German-speaking areas in Pennsylvania and other ethnic enclave holdouts, square rule was widely adopted and became the rule of the land. An excerpt from the time of its introduction in southern New York (from *The History of Delaware County*, by W. W. Munsell, 1880) tells about the rapid transition there:

> Judd Raymond was the first regular carpenter and worked by the "scribe rule" several years. He settled in 1792 on the farm north of Squire Eells's — now owned by Thompson. He was followed by Thomas Dennis and Mr. Sweezer in 1795, who also worked by the "scribe rule." George Dennis came in 1804 and the "square rule" work was soon instituted. Benjamin B. Eells, a son of John Eells, was the first to learn, and in 1805, when only eighteen years of age, he framed and raised the first barn by that rule; it was on Mt. Pleasant, and people assembled from far and near to see the lad's failure, but their pleasure and astonishment were unbounded on seeing the great improvements of the age. The town has now scores of good workmen.

THE PERFECT TIMBER WITHIN

In square rule layout, lines are snapped on the faces of a hewn timber to describe the joinery plane and the ideal timber size. The framing square is then applied to these lines when laying out the work.

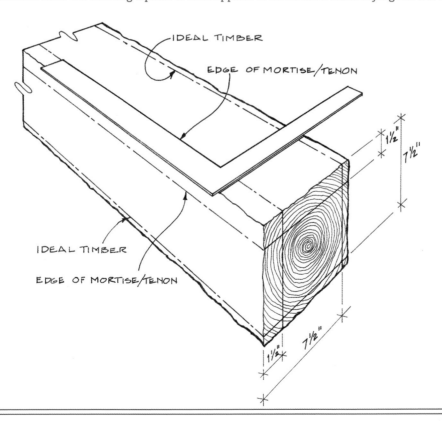

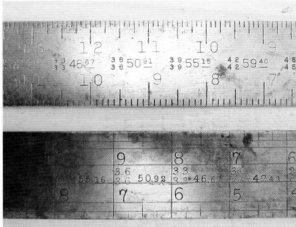

In the square rule, the carpenter used the framing square (hence the name) as his pattern for mortises, tenons, and pin holes. The framing square, still in use today, was fairly standardized by the early nineteenth century, with a 2-inch-wide "blade" that is 24 inches long, and a 1½-inch-wide "tongue" that is 16 inches long. The 2-inch width of the blade was used to lay out a 2-inch mortise or tenon, typical for softwood framing. The 1½-inch-wide tongue was similarly used for hardwood framing. The tongue or blade of the square was also used to check the finished mortise for size. Pin hole centers were typically either 1½ inches or 2 inches off the face of a mortise or tenon. Yes, the pin hole was laid out with the square on both the mortise and tenon! When boring, the workman offset the pin hole on the tenon by the appropriate amount (⅛ to ⅜ inch) closer to the tenon's shoulder to provide the drawbore.

Many squares had a brace table stamped into them, providing the diagonal lengths of given brace sizes and eliminating the need for some mathematical figuring on the carpenter's part. A sizable square-rule barn frame could have a hundred or more diagonal braces of all the same size, all interchangeable. Therefore, the brace table was a very convenient feature. Some framing squares also had rafter tables providing the length of run (horizontal dimension) per foot of rise (vertical dimension) for different roof pitches — another handy feature.

Compared to the scribe rule, the square rule required a lot of skills in measuring, adding, and subtracting. However, if the workman used the square, got his math right, and could cut to a line, the frame would be raised square, plumb, and true with all its joints snug — and in considerably less time.

Shifting Styles and the Rise of the "Balloon" Frame

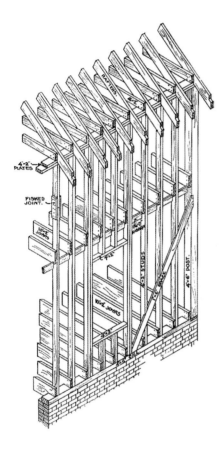

Balloon framing utilizes 2×4 studs rising the full height of the walls. Second-floor joists are nailed to the sides of the studs and further supported by a 1×4 ribbon let into the studs. For racking resistance, 1×4 diagonal corner braces are let into the studs.

THE UPPER ECHELON OF SOCIETY has always been concerned with style, but with the advent of pattern books in the late eighteenth century, style consciousness was brought to the forefront of popular building design. Books such as those by Robert Morris, Isaac Ware, Abraham Swan, and Batty Langley in England, along with America's own Asher Benjamin, illustrated properly proportioned and classically ornamented structures that the public would aspire to. Most included proportioning for the Tuscan, Doric, Ionic, Corinthian, and Composite orders of columns and pediments to create classically based designs, as well as treatises on geometry and mathematics. Much like the glossy home magazines of today, pattern books introduced the new styles to the public.

After starting out with the English styles of Elizabethan and Jacobean, and later Georgian, America developed its own post-independence styles of Federal, Greek Revival, Italianate, and Gothic Revival. It seems that every generation wants a new style, and house construction and floor plans change accordingly. Owners of existing houses were often shamed into updating their house's exterior with the latest style. Some houses show evidence of multiple periods of adornment.

After about 1750, exposed timber frames in houses became less fashionable, and the frames were typically boxed in with beaded boarding or completely concealed in the walls. Plaster, moldings, trim, murals, and wallpapers were the desired interior features, though the timber frame still provided support.

Three developments of the industrial age effectively removed timber framing from the mainstream home-building industry. The primary innovation was the development of balloon framing by George Washington Snow in Chicago in 1832. This lightweight (hence the term "balloon") building method uses small — typically 2-inch by 4-inch — mill-sawn uprights (wall studs) spaced closely together and extending a full two stories, and joists and rafters, also 2 inches thick, secured to the uprights with nails instead of wooden joinery.

This framing was much faster to erect and required much less skill than timber framing. Rather than apprentice to a master carpenter for years, workers were trained in a matter of days. The speed of the method and its quick learning curve made it indispensable for rapid rebuilding of cities after a fire or for creating Western mining and railroad towns overnight.

At the same time, timber framing was beginning to be described in a derogatory manner:

> Thus you will have a frame without a tenon or mortice, or brace, and yet it is far cheaper, and incalculably stronger when finished, than though it was composed of timbers ten inches square, with a thousand auger holes and a hundred days work with the chisel and adze, making holes and pins to fill them.
>
> *(From a lecture by Solon Robinson in New York City in 1855)*

The claim that balloon framing is stronger than timber framing is not true. It should have read that balloon framing is "incalculably stronger than one would think for such small pieces."

The second innovation that contributed to the shift in construction styles was the machine-produced nail. Balloon framing used a lot of nails, so a source of inexpensive and standardized nails became essential. Prior

Styles of Early American Homes

The 1863 Parson Capen house in Topsfield, Massachusetts is a prime example of an Elizabethan-style house. Typical of the style, it has tiny leaded glass windows, jetties at the floor levels, a steep roof pitch, and exterior timber embellishments.

The 1768 Joseph Barnard house in Deerfield, Massachusetts, is a Georgian-style house with a hipped gambrel roof.

to about 1800, blacksmiths forged each nail by hand, as they had for eons. Because nails were expensive, building construction systems that used fewer of them had an advantage. The advent of machines that produced cut nails changed that. At first, tapered nail shanks were sheared out of metal sheet stock and headed by hand. Later, heading became part of the machine process. By the 1890s, a further improvement, the wire nail with its round shank and head, became the norm for construction.

The third innovation that finally propelled the 2×4 building method into the mainstream was the circular sawmill. The older, water-powered, up-and-down type of sawmill was the norm until the mid-nineteenth century. Though circular saws were in use earlier for re-sawing pieces into smaller sections like lath and trim, they were essentially large table saws and not for sawing full-size logs. The development of water turbines and the rise of steam power enabled these much faster mills to become the standard by the 1860s. The steam-powered mills didn't require a stream, dam, and millpond arrangement — they were essentially portable and were set up in the forest.

Federal-style houses will often have the same symmetry in the façade as the Georgian style, but chimneys are typically in the gable ends.

In the Greek Revival style, the house alludes to a Greek temple with the gable facing forward. The façade is typically three units wide with the door to one side. There may be pilasters and even columns on the more upscale versions.

With the combination of balloon framing, inexpensive nails, and efficient sawmilling, a craft that remained a standard for 7,000 years was relegated to a niche in the building industry, where it remains to this day. Balloon framing eventually developed into platform framing, or "stick framing," as it is commonly called today. Studs no longer rise the full height of the house but are capped off by floor framing at each level.

Timber framing dominated in house construction at least through the Greek Revival period (mid-nineteenth century) and until much later in church construction, where the framing was either concealed or exposed on the interior. It was also popular for barn and mill construction into the early twentieth century because, as it turned out, 2×4s weren't substantial enough to withstand the heavier loads in these types of building.

In this great late-nineteenth- or early-twentieth-century construction shot, an early-nineteenth-century timber-framed house has been moved intact from another location and sits up on cribbing. Its gable wall has been stripped to expose the timber framing and the lath and plaster. Behind the man standing in the foreground is a capstan (missing its drum) that was probably used to move the house. In the right rear is a late-nineteenth-century balloon-framed house probably in its original location. In the foreground, carpenters are balloon framing a large section with multiple doorways. The joists will bear on the horizontal "ribbon" at the second floor. Note also the diagonal board sheathing that gives the building rigidity. Masons are also at work building a fireplace and chimney at right, which accounts for some of the untidiness of the job site. While the location is unknown, it appears that the building assemblage will be some sort of lodge, country club, or inn.

Even in residential construction, timber sills were still being used in conjunction with balloon framing in some places into the early twentieth century. Before the advent of the poured concrete foundation, stone foundations were the norm. The sill timber not only defined and supported the house's structure, it tied the stonework together along the top, distributing the building's weight more uniformly over the foundation. Setting and leveling a 2×4 sill on stonework would be a trying affair, so builders often used sill timbers joined by mortises and tenons at the base of 2×4 walls. For a solid poured concrete wall, however, the smaller pieces of lumber work just fine and have become the norm.

Later, timber framing enjoyed a bit of a revival during the Tudor, Gothic Revival, and Arts and Crafts periods, in which exposed and nicely finished timbers were once again fashionable. Timber-frame details, such as arched braced collars and hammer-beamed roofs, were often seen in churches and chapels, though they might have been built up with smaller members and have bolted connections rather than mortise and tenons.

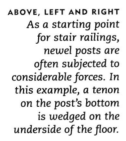

Where to Look: Barns, Churches, Mills, and Bridges

Covered bridges are rare today, but in the past they were very common. The earlier types were timber-framed, and the roofs protected the wood from the elements. An open wooden bridge could be expected to last a decade or so. A covered bridge, with occasional maintenance, could last a century or more. Most of the surviving bridges are of a late style that may use large members but used dowels, bolts, or nails instead of the wooden joinery of the earlier versions.

MOST PEOPLE ARE FAMILIAR with the craft of timber framing through seeing old barns, where the framing is exposed. Because it is covered up in houses after about 1750, it is much less understood in this context. The cellar and the attic are often the only places where it is open to view. But if you look closely, you can see timber framing in a few other elements. Many of the windows of houses were framed with 3- and 4-inch-thick timbers that were mortise-and-tenoned at the corners. Also, because barn doors are large and need to handle substantial wind loads, many were framed with small timbers, called *scantlings*. Timber joinery can be utilized even in a house's finish work: many stair newel posts are cut from a timber-sized chunk and may have wooden joinery to secure them to the floor framing below.

Churches are typically timber-framed, but most of the framing is covered up. One has to climb up steep stairs and ladders to reach the attics and spires where there is often some very impressive timber framing. Because of the wide-open sanctuary space below, the roofs are usually trussed and typically contain some big, long, elaborately joined timbers. In fact, I'd have to say the most impressive historic timber framing in America resides in church attics. Unfortunately, poor lighting and the difficulty of getting around up there make it hard to get a good look. There may be loose insulation covering the ceiling framing and rarely any floorboards. Stepping between the joists could land a poor fellow on the sanctuary floor 30 feet below. Very few parishioners get up into the attics and spires of their churches.

Mills of all types are timber-framed, as were covered bridges. Less common timber-framed structure types included forts, towers, lighthouses, railroad structures, and lock gates for canals. Even ephemeral or temporary constructions, such as hoisting devices, scaffolding, falsework, centering for stone arches, and warfare siege engines, would be timber-framed.

UNCOVERING THE FRAME OF A CHURCH

A truss is an assembly of pieces configured to span greater distances than a single member to support floor and roof loads. It usually involves triangulation of components and often fills the roof space. This particular truss is a scissors truss. It supports the roof and arched ceiling of St. Paul's Episcopal Church in Windsor, Vermont. Built in 1822, its white pine trusses span 48 feet and bear on brick walls. A scissors truss can be an advantage because it does not require a tie beam the full width of the structure and the ceiling can rise above the walls. Its disadvantage is that because of the higher than normal stresses, the timber joinery must be cut to close tolerances to prevent spreading and sagging.

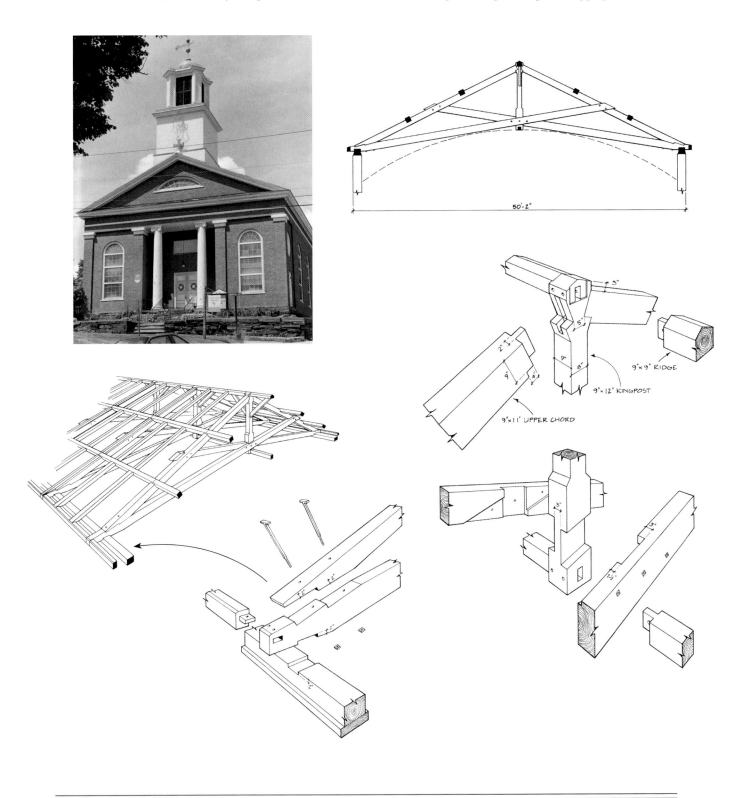

50'-2"

9"x9" RIDGE

9"x12" KINGPOST

9"x11" UPPER CHORD

TAKING APART A STEEPLE

The Middlebury, Vermont, Congregational Church (1806–1809), with its
135-foot-high spire, is a beautiful example of Federal design.
The steeple comprises two square stages and two octagonal ones,
all decorated with pilasters and arches. However, to a timber-framing enthusiast,
the real gem is the superstructure within the steeple.
To prevent the steeple from toppling over in the wind or lifting off in a hurricane,
historic builders would often make use of a technique referred to as telescoping.
Rather than one steeple stage being simply stacked upon the other,
each stage extends down into the previous one at least as far as its own height.
The base of the steeple bears upon the attic floor framing in the rear and extends to the foundation
in the front two posts. The mast, to which the spire rafters are attached, extends down
through three stages. There is practically a forest of timber in this steeple!

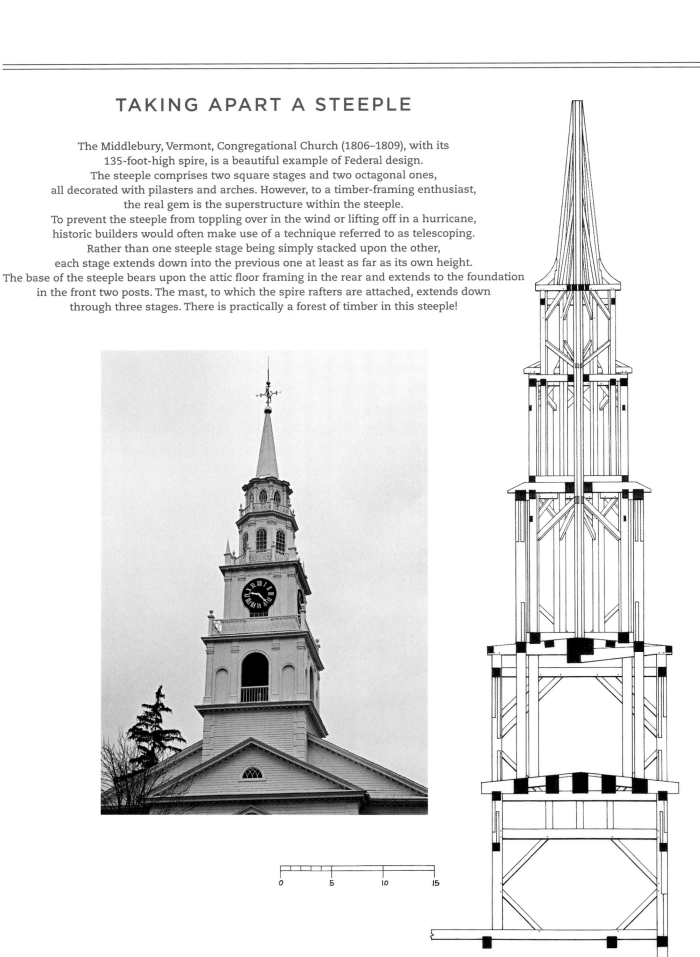

0   5   10   15

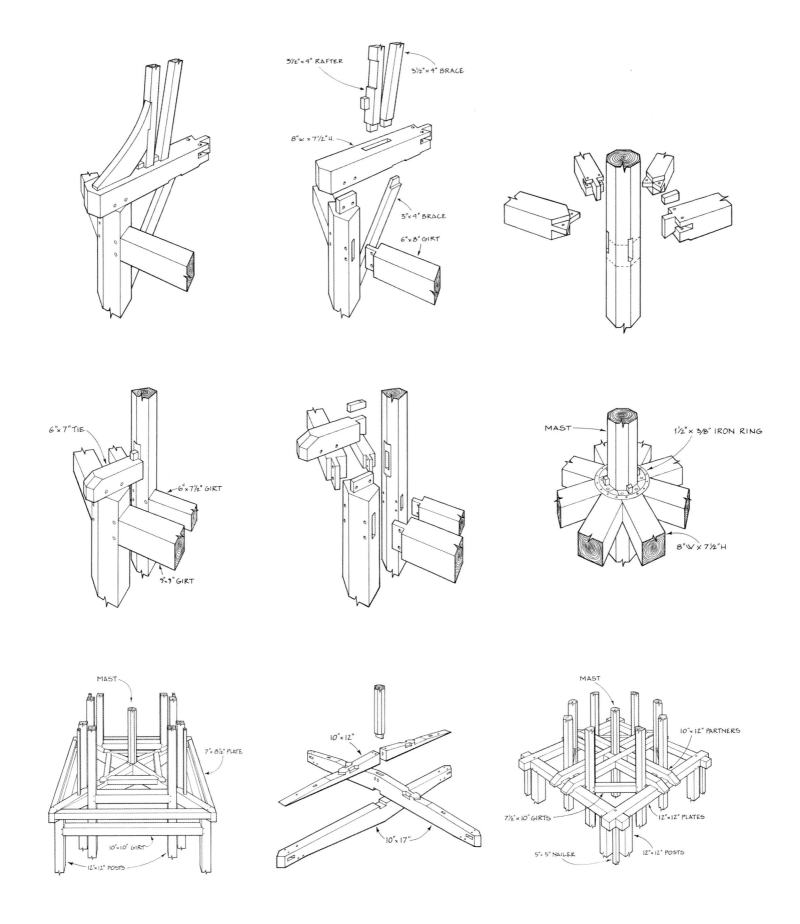

3½" × 4" RAFTER

3½" × 4" BRACE

8" W. × 7½" H.

3" × 4" BRACE

6" × 8" GIRT

6" × 7" TIE

6" × 7½" GIRT

9" × 9" GIRT

MAST

1½" × 3/8" IRON RING

8" W × 7½" H

MAST

7" × 8½" PLATE

10" × 10" GIRT

12" × 12" POSTS

10" × 12"

10" × 17"

MAST

10" × 12" PARTNERS

7½" × 10" GIRTS

12" × 12" PLATES

5" × 5" NAILER

12" × 12" POSTS

The Back-to-the-Land Movement and Timber Framing's Revival

Counterculture, hippies, and the back-to-the-land movement produced some unique examples of the building craft.

IN THE LATE 1960S AND EARLY '70S, a growing number of young people were becoming dissatisfied with the status quo. They were not only tired of the Vietnam war, but also of the machinery of corporations, big business, and government interference. Plus, there was rock music, psychedelic drugs, television, and the space age and all the new technology associated with it. Many wanted a simpler, more independent life, free of modern trappings. Thus, the back-to-the-land movement was born.

Back-to-the-landers wanted to live off the land, grow or hunt for their own food, and commune with nature. They wanted to build their own houses, too. True to their creed of living simply, they built houses from found or salvaged materials, using bottles and tires or even driftwood. They built geodesic domes, yurts, tipis, tents, and inflatable houses. They converted old school buses, churches, barns, and silos into homes. They built underground or up in trees. They built pioneer-type dwellings of adobe, earth, sod, stone, and log. And, of course, they built timber frames. Yes, timber-framing your own building was a liberating experience: trees shaped by axes into timbers, timbers connected by mortise-and-tenon joinery, joints held tight by carved wooden pins driven with wooden mallets. And raising the frame, that was a *happening!* Your friends all came to help lift and to help celebrate. It was unforgettable.

These were formative times. And while some were overcome by the times and languished, unable to move beyond the hippie stage, others developed their building skills into more professional careers using adobe, sod, or logs as well as timber framing. They worked with the building industry and building codes and researched historic precedents, and soon these revival methods became more mainstream and accepted by the public. By the '70s, many individuals and small firms were building new timber-framed homes or repairing old ones. In 1985, a conference took place at Hancock Shaker Village in western Massachusetts. Two hundred interested timber framers assembled from all over the United States and Canada. Out of this meeting was formed the Timber Framers Guild of North America. Since that time, the free sharing of information between members through conferences, community raisings, and publications has brought the craft to new heights.

The Timber Framers Guild has been building community since its inception. These two house frames were raised for Habitat for Humanity in 1989 in Hanover, Pennsylvania, as part of an annual conference. Guild members from as far away as England participated, and each brought a piece they had cut to assemble into the frames.

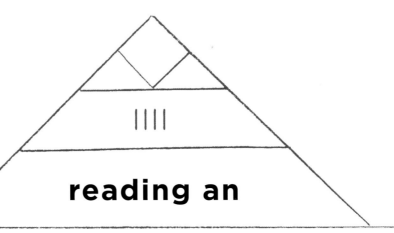

reading an

OLD BUILDING

WHEN ENTERING AN OLD timber-framed structure there is much for the enthusiast or craftsman to take in. I look at the arrangement and size of the members and the marks of the tools that turned them from logs to timbers. I can determine what species were used and whether the trees came from original forest or second growth. I can examine the joinery and the pins securing it as well as the various tool marks that offer clues to the fabrication process. The carpentry also shows evidence of the building's history: how it was used and how it was modified over its lifetime. In larger and more complex structures, there can be so much to take in that it can be overwhelming.

From the Outside In

WHEN I APPROACH A BUILDING from the outside, I can often glean information on what may lie within. Overall shape and proportion, roof pitch, chimney position, fenestration (windows and doors), material choices, siding, cornice, trim, foundation type, and even the building's position in the landscape can provide clues. However, many buildings have been modified over their lifetimes, moved from another site, raised up, added to, or greatly changed in appearance. Over the years, I have had some surprises when the outside appearance didn't match what I found within.

When I do step inside, if it is a barn or a building with an exposed framing system, I'll first look at the timber work. Is it hewn or sawn, square rule or scribe? What is the configuration of the bents? What joinery is used? Is the wood old growth or second growth, and what species were used? Are the pins riven or turned on a lathe? Are they different sizes? Invariably, even in the simplest of buildings, I will spot something I haven't seen before. It may be a prominent feature, or it may be quite subtle: a tool mark or a slight variation in the joinery or pinning.

When I enter a building with a concealed timber frame, say, a nineteenth-century house, there is still much information to take in. The interior finish — doors, mouldings, hardware, flooring, and stairs — all provide clues to the building's age and how it was used over its lifetime, as well as how it was changed. If I get a chance to see some timber work in the attic or basement, most of my questions are answered. But of course there are buildings that are enigmas, that release their story slowly and not on the first visit. I like a good mystery and the chance to postulate on what I find.

In larger, more complex buildings such as the 1821 Mount Lebanon (New York) Shaker meetinghouse, there is so much to take in when seeing the attic for the first time that it is truly overwhelming. With an arched roof and curved ceiling below, the timber framing is exemplary. For example, the 65-foot tie beams were first hewn square, then sawed down their length in order to notch the radiating struts through them.

MEASURED DRAWINGS

Some elements of a building's design and layout can be gleaned only by taking measurements and subsequently drawing it up to scale. While photographs can be taken quickly and can adequately record a lot of information, recording with measured drawings helps you to *understand* the building more completely. A scaled drawing shows proportions of the elements and relationships between them. In order to draw something, you must examine it closely, take some measurements, and look at it from different angles. As architects, we are taught to draw; and by drawing, we understand.

ABOVE *Attic floor framing in an eighteenth-century house*

BELOW *A drawn plan of the same floor structure*

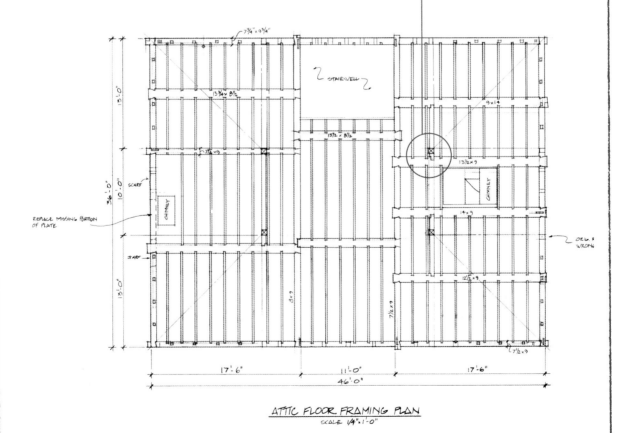

ATTIC FLOOR FRAMING PLAN
SCALE 1/4"=1'-0"

Timber Species

THERE ARE MANY WHO CLAIM to be knowledgeable about old wood, who sing its praises and swear by it, but there are relatively few who can properly identify the species of an old timber. Most people cannot identify a new timber, much less an old weathered or dust-covered one. Consequently, there is much misinformation out there about the species used in old buildings.

Most so-called experts will guess that it's oak or chestnut. And in certain areas of the United States they would have a great likelihood of being right. For example, in southern New England, where the oak-hickory forest type now predominates, and did historically, oak was the first choice of builders for a timber frame. The early settlers hailed mostly from the British Isles, and their preferred wood in their homeland was also oak (*Quercus robur* and *Quercus petraea*), so it was a natural choice. Of the multiple varieties of oaks that grew here, they preferred trees from the white oak group because of their rot resistance. Trees from the red oak group decay surprisingly quickly. The settlers' second choice was likely American chestnut, as it was also rot resistant, though lighter and softer than oak. It also grew straighter, taller, and faster than the white oak, making it more common in second-generation buildings as the oaks were being depleted.

If, however, one could record the species of all the surviving old timber-framed buildings of New England as a whole, I don't think oak or chestnut would even make it into the top two places. Based on a rough tally of the observations of builders throughout New England involved in the repair of old structures, I'd say it is likely that American beech and red spruce would predominate. Most of Maine, New Hampshire, and Vermont, as well as the higher elevations in western Massachusetts and Connecticut, are characterized by forest types that differ from the oak-hickory of southern New England. These types range from transitional forests of hardwood, white pine, and hemlock to northern hardwood forest (beech, yellow birch, sugar maple) and up to northern coniferous forest (red spruce and balsam fir).

In parts of coastal New England and in interior river valleys where sandy, well-drained soils are found, pitch pine forests were predominant. Pitch pine is in the southern yellow pine group that happens to grow in much of New England and New York. The southern yellow pines also include loblolly pine, shortleaf pine, longleaf pine, and slash pine. Once the trees are harvested into wood products, it is impossible to separate the individual species. Though there are few pitch pines of any quality growing today, in the past they were quite respectable timber trees and were abundant. The wood is extremely strong, heavy, and durable. There are 12 × 24-inch pitch pine anchor beams 30 feet long surviving in Dutch barns in New York State.

TOP *This pitch pine anchor beam end was salvaged from a barn on the banks of the Mohawk River in Rotterdam Junction, New York. It measures 12 × 24 inches.*

BOTTOM *I once found rock elm timber cutoffs used as blocks to support interior cross sill girders in a Dutch barn in New York State. The cutoff was from the barn frame and had sat on the ground for over 200 years and was still sound!*

If we took all the historical timber frames east of the Mississippi River, white oak might indeed top the list. However, most frames are composed of more than one species. And there are often some oddball species tossed into the mix. In the first settlements, what was typically used were the trees standing in the owner's woodland. The owner (farmer) most likely cut the trees himself and either hewed them or brought them to a local sawmill. While the posts might be oak, the plates and purlin plates might be pine or basswood. The rafters could be yellow birch or ironwood!

Surprisingly, even elm was used. Anyone who has cursed when splitting elm for firewood has experienced its *interlocking grain*, in which each year's growth spirals in a different direction. I have seen huge swing beams in barns nearly 2 feet through and 30 feet long, hand hewn from elm. Rock elm is the preferred elm, as it is heavier, stronger, and, most importantly, very rot resistant.

Black ash (*Fraxinus nigra*) is a wood hardly ever mentioned but common nonetheless for timber frames. It grows in lowland woods with continually wet soils and produces tall, straight, clear trunks. Today it is prized by traditional splint basket makers, who use the beautiful brownish wood with a silvery sheen to produce their splints.

The frame of the education center at the Mountain Top Arboretum in Tannersville, New York, is made from 21 different species of trees harvested from arboretum land.

From Tree to Timber: Conversion Methods

THERE ARE THREE PRIMARY METHODS of converting a round log into a more usable rectangular shape: hewing, riving, and sawing. Historic timber-framed buildings may have been converted by only one method or by a combination of methods. Conversion methods can give us clues to a building's age, how it was modified, or the distance to the nearest sawmill. Each method leaves telltale marks, clues for someone piecing together the story of a building.

hand hewing

Hewing is perhaps the oldest conversion method, as it can be accomplished with the most primitive of tools: an axe. Even the earliest axes of stone could hew timbers. With the advent of bronze and then iron, hewing advanced considerably, but even into more modern times, the tool remained simple: a hand-forged iron axe was still a relatively easy tool to make and sharpen. Hewing is not a highly skilled art, and in former times every farm boy learned to use an axe as soon as he was able. In teaching my timber framing workshops, I'll often demonstrate hand hewing a timber from a log. Students learn the technique quickly. By the end of the first day, many can do a respectable job of it. As with any craft or trade, some have a natural inclination for it and will excel in it. I have found that many people will take great pleasure in shaping wood with their own hands and marvel at the thought that a tree can be shaped into a timber for their use with only an axe.

Though one may assume that hewing as a craft is archaic, and the broad axe only suitable today for display above the fireplace, it is far from a dead craft. It appeals to the frugality and the desire for empowerment in all of us. It probably always has. With simple, inexpensive tools and very little out-of-pocket cost, one can, with satisfaction, create the bones of one's house.

Timber-frame builder Neil Godden has begun squaring up a freshly felled eastern white pine log into a 10×10 timber, right in the forest.

Hewing

LINING

Hewing is accomplished by first "lining" the log to describe the member being hewn from it. Straight chalk lines are snapped on its surface to describe flat, vertical planes within the round log.

Next, standing on top of the log, the hewer scores the wood to the line with a regular felling axe every few inches, and the wood between the cuts pops off. In this way, the waste material is removed to prepare the surface of the timber for smoothing.

On larger logs where a great deal of material must be removed, the hewer makes use of a time-saving technique called "juggling." A V-shaped cut is made to the line at wider intervals from 1 to 3 feet. The big chunks between these notches can then be split off with the axe. The surface is then re-scored to the line as needed. This not only saves time, it also provides nice take-home chunks of firewood.

On pine and spruce logs, these notches are made at the knot whorls (tiers of branches). Removing the knots makes the splitting off of the waste chunks much easier. If the knots are tiny or not visible on the surface, juggling can be done at a regular spacing. The V-notches from the juggling process are often still visible when the hewing is done. In one old barn I viewed, the juggling cuts were at exactly 16 inches on center and not aligned with the knot whorls. I suspect the hewer had a wood-burning stove that could handle chunks up to 16 inches long, and he was perhaps a bit exacting about the length of his firewood!

HEWING WITH THE BROAD AXE

Next comes the broad axe, a wider-bladed, shorter-handled, heavier axe that hews to the line. Kneeling with one leg on the log and the other leg on the ground, the hewer starts at the top of the log and hews backward toward the butt end. Diagonal, slicing strokes are used to pare away the material to the line.

A good hewer can leave a relatively smooth, flat surface that needs no cleaning up with a plane.

THE TOTAL
TIME IT TOOK
TO HEW THIS
TIMBER WAS
2.5 HOURS.

After the two vertical faces are hewn, the piece is turned and the process repeated for the remaining two faces. To save time and effort, some pieces in a frame were not hewn square but rather given a flat face only where necessary. Floor joists, wall girts, and rafters are often hewn on only one face to provide a flat surface to accept floor, wall, or roof boards (see page 77). Usually the bark was removed so as not to attract wood-boring insects. Another way to save hewing time was to follow the natural taper of the log, thereby removing less material. Tapered timbers were commonly used for rafters, posts, and occasionally plates and ties.

Riving

Riving begins by halving a freshly felled, straight-grained log using wedges.

The halved sections are further split using the froe and wedges.

This 3 × 5-inch section can be further worked with a broad axe or hatchet as necessary for its intended use.

riving

Riving is a process of cleaving or splitting wood along its grain. Woodworkers have always known that if you are working with clear, straight-grained, high-quality material, riving is a very fast and efficient way of roughing out stock, and the end result is a great-quality piece. The original old-growth forests here were composed of many tall, straight, branch-free trunks, ideal for riving. To provide land for agriculture and pasture for grazing animals, great swaths of forest had to be cleared quickly. With so much timber, a woodworker could choose the best, and the rest was burned.

The riving technique was especially useful for smaller members, like studs, braces, rafters, and joists, where multiple pieces could be riven from a larger, clear log in a very short time. The rules of riving are simple and few:

» First, start with the best material — green (freshly felled), straight-grained, branch-free logs of species that split readily. The best part of the tree is the lower end where the branches have long ago died off and clear wood has grown over, but above the stump flare. The first 5 or 6 feet often have irregular grain with a lot of taper. When trees are young saplings, the buds are often eaten by creatures such as deer, moose, and rabbits. When a bud is chewed off by a grazer, the tree reshoots below it, often by forking and creating multiple stems. It can take several years, if not decades, for a sapling to finally grow its leader above the browsing height of animals. A tapered, gnarly-barked butt section will not be the best candidate for riving.

» Second, for a split to follow the grain exactly, there should be roughly equal wood on each side of the split. Otherwise, the split will run to the thinner side and possibly run out completely. Therefore, riving is best when each split halves the piece.

» Third, always start a split from the bottom end of the log, especially when there is noticeable taper in the log. The grain in a tapered log is parallel with the outer surface, not the pith, or center. Growth rings are actually elongated cones that come to a point representing the tree's height in that year. When riving a log in half from the bottom, the split will follow the pith rather than the grain. When riving from the top, the split will follow the grain, parallel with the bark, and eventually start to run out.

To start a split, an axe or froe (also spelled *frow*) is driven into the wood. As the split opens, wedges of wood or iron are inserted farther down the log to continue the split. In easier-splitting species like red oak, the log will practically "pop" apart with little wedging.

Riving roughs out a member, but it still needs some shaping to be useable. Traditionally, once the pieces were split they were often lightly scored and hewn flat with an axe. Generally, riving was utilized for shorter members under about 8 feet long, but I have seen some riven rafters at 20 feet and beams of about 16 feet. Besides timbers, riving was used for making the pins, clapboards, and shingles, and occasionally other building elements.

RIGHT *This gable end stud in a house attic was clearly riven and shows no further dressing on the interior surface.*

BELOW *The trunk of a tree can be seen as a series of stacked cones, as shown (and exaggerated for clarity). If the split is started from the top (center), it will follow the grain, running out. If started from the butt (right), it will be more likely to continue down the center.*

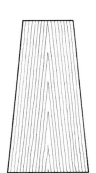 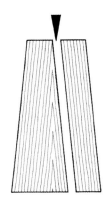 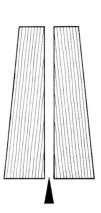

Sawing by Hand

ABOVE, LEFT *Wheelwright and blacksmith Sidney Franklin saws through a 50-year-old oak, held firm by iron dogs on the beams over his saw pit in Essex, England.*

ABOVE, RIGHT *This illustration from Diderot & D'Alembert's Encyclopédie of 1762 shows two-man ripsawing. The timber is elevated on wooden trestles rather than worked over a pit.*

RIGHT *Seesawing leaves a telltale triangular riven surface in the center of the piece. This English cruck blade was seesawed from a poplar log. The red lines indicate the end of sawing from each direction before it split apart.*

sawing

The third method of conversion is sawing, either by hand or with a mill. Sawing works regardless of the grain, knots, or twists and produces flat-surfaced members. In fact, sawing can make timbers from logs that are impossible to hew or rive. This can be a disadvantage, however, because a sawyer can saw out an acceptable-looking timber from a really poor-quality log, possibly fooling the novice.

In early American settlements that were far from a water-powered saw-mill or in areas where water power was unsuitable, timbers would be sawn by hand using a 6- or 7-foot, two-person ripsaw. The log was either set up over a hole dug in the ground (pit sawing) or on a trestle 6 feet above the ground. One person was positioned on the top of the trestle to guide the cut along a snapped line and lift the saw up. The second person, called the pit man, was positioned on the ground below and pulled the saw down on the cutting stroke.

Another method was *seesawing*. One end of the log rested on the ground while the middle was supported by a sawhorse. The end in the air would be sawed toward the center over the sawhorse. Then the log was reversed so the end formerly on the ground was up in the air, and the sawyers again worked from the high end toward the center. When the cuts met, the sections would fall off. Ripsawing by hand was a laborious but effective method of sawing out timbers, planks, and boards in the absence of a water- or wind-powered sawmill.

In America, sawmills were set up almost immediately following the earliest settlements. One was set up on the James River in the Virginia colony in 1611. Along with gristmills, sawmills were deemed essential for a rapid colonization of the New World.

The sawmill was simply a mechanically powered version of ripsawing. A long, straight blade was mounted in a frame that moved up and down as

Early Sawmills

The water-powered sawmill was essential for the settlement of towns, allowing colonists to turn the vast, forested wilderness into useful building products. This reconstructed version operates at Old Sturbridge Village, in Sturbridge, Massachusetts. Early mills used an up-and-down type of saw that was quite slow by today's standards.

The saw could not cut all the way to the end of the log because the log was held onto the saw carriage by iron dogs driven into one end, with the saw stopping just shy of the dogs. The slab or board would be snapped off the cant (partially sawn log), leaving a telltale riven section on the last few inches of the board.

the log was advanced into it. Water spilling onto a water wheel turned shafts with gears to power the blade and the carriage that advanced the log. The power was controlled by starting or stopping the flow of water to the wheel. Compared to the later circular sawmills, these up-and-down mills were slow; each pass through the saw would take several minutes. They were also limited in their capacity, usually 16 feet in length, so longer timbers still had to be hewn.

By the Civil War, circular sawmills had become the norm and were often powered by turbines rather than water wheels, or by steam engines, which made the sawmill somewhat portable. Though the circular saw had appeared decades before, it originally had been used primarily for re-sawing timbers or planks into smaller pieces, such as laths for plastering. The circular sawmill was considerably faster than the up-and-down saw, taking only seconds for each pass instead of minutes.

But, amazingly, hand hewing persisted well into the early twentieth century. While the circular sawmill was fast and convenient, a farmer still had to *pay* for the sawing work, and perhaps draw his own logs to the mill. Alternatively, he could fell and hew timbers right on his own property without any monetary outlay. Hewing also was not subject to constraints on log length, while most mills could cut only up to 16 or 20 feet.

TOP *A circular sawmill in Alabama in 1935*

BOTTOM *The Canadian Forestry Battalion cutting timber with a circular saw in a steam sawmill at Virginia Water, Surrey*

A ROOF PITCH MYSTERY, SOLVED

On one early-nineteenth-century English barn restoration that I was involved with, the rafters were sawn rather than hewn, and the roof pitch was a flatter 7/12 instead of the more common 9/12 pitch. This anomaly was easily accounted for when I discovered that the property had once had a working up-and-down sawmill on it. The dam was still intact, not far from the barn. An owner constructing his own barn would be hesitant to hew out timbers with an axe if he owned and operated a sawmill (still true today). However, it appears that his mill could handle logs only up to 16 feet, so he narrowed the barn to 26 feet (from 30) and reduced the roof pitch to a 7/12 so that he would be able to use 16-foot sawn rafters.

David E. Lanoue Inc.

ABOVE Loggers build rafts from felled trees to float them down the Susquehanna River in Pennsylvania, on the route of the Erie Railroad.

RIGHT This beautiful nineteenth-century swing-beam barn originally stood in Caledonia, Ontario, near the Grand River. It is being set up in-shop following repairs. Many of the timbers in this 30 × 54-foot pine barn have evidence of rafting.

Rafting

IN AREAS NEAR LARGER NAVIGABLE RIVERS, timber was often transported through a process called rafting. In the upper reaches of the river's watershed where white pine was abundant, trees were felled and hewn to consistent sizes, often 12 inches square and 40 or 50 feet long. These were dragged by oxen to the frozen river where they were assembled into cribs of timbers about 25 feet wide. In spring, the cribs were manned by raftsmen and floated down the streams to larger rivers, where as many as 200 cribs were joined into huge rafts that could hold as much as 80,000 to 120,000 cubic feet of timber.

Since the river journey could last months, the raft was outfitted with cooking and sleeping accommodations for the drivers. Raftsmen would man tillers to steer the huge craft and would tie up the raft to the bank at night. For crossing narrows and rapids, the raft might need to be broken down into smaller sections and then reassembled on the other side. After floating downstream to more settled areas or seaports, the timber could be sold to waiting ships for export, or it could be re-sawn at a riverside mill into smaller sizes for local building use. Some of the larger rivers involved in this rafting of squared timber were the Hudson, Mississippi, Ohio, Allegheny, Susquehanna, Delaware, and Ottawa.

With close inspection, it is possible to find clues pointing to the use of rafting. Timbers hewn on some faces and sawn on others indicate that timbers were first hewed with axes and then sawed into smaller components at a mill. Another clue is a frame composed of nearly all 12 × 12 pine timbers.

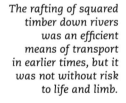

The rafting of squared timber down rivers was an efficient means of transport in earlier times, but it was not without risk to life and limb.

Rafting Clues

Sawn boards may bear the remnants of the pins that were used to assemble the rafts. This board was found in a 1760s house in West Springfield, Massachusetts, and probably represents a rafting tradition on the Connecticut River.

This ash pin was chopped off flush with its hewn timber. It was from a barn in Caledonia, Ontario, near the Grand River.

This barn door header was sawn from a hewn timber that was likely rafted. The sawing bisected a pair of holes, one of which contained two pins wedged against each other. This setup could have been used to secure a rudder to the raft of timber.

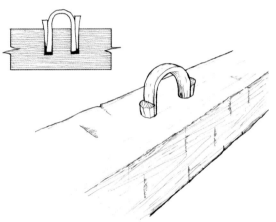

This sketch shows how the timber at left might have been set up to hold a rudder for rafting.

Also, where you would expect to find rectangular dimensions for sills and plates, with rafted timber these pieces are square.

The presence of chopped-off pins is yet another clue. To assemble timbers into a raft, smaller members were laid crossways and secured to the larger timbers below. Rather than use costly iron spikes, the loggers used wooden pins fashioned on-site. Though the hardwood pins were probably set into holes pre-bored in the smaller crossways members, they were sharpened to flattish points and driven into the 12 × 12 timbers 2 to 4 inches deep. When the rafts were disassembled, these pins were difficult to pull out, so instead they were chopped off flush. Their severed ends still remain visible on the timbers. If the timbers were subsequently sawn into boards, the sawed-off peg tips may be visible.

Fire-Scarred and Dead Standing Timber

SOMETIMES TIMBERS IN OLD STRUCTURES may have charring on their waney edges only, indicating that an already charred log was squared up into a timber. The *wane* is the rounded surface of the wood just under the bark. I don't think the charred edge is an indication of some process used to expedite the work but rather just a matter of using material that was on hand. Standing trees charred by fires from clearing land, or perhaps from natural causes, can still be used for building. Fire was a tool used to clear the undergrowth in forests so that farmers could plant between trees rather than wait to remove them and pull their stumps. It shouldn't be surprising that first-generation buildings might have some charred timbers in them.

A standing dead tree less than a year or so dead is also fine for a timber. If decay has set into the sapwood portions, it can still be sawn or hewn, and the decayed portions will be removed in the process. I occasionally find white pine timbers with holes from the pine sawyer beetle. Since these beetles attack only killed trees or freshly cut logs with bark still on them, and since they take months to do their work, such timbers were probably cut from dead standing trees or logs that sat on the ground unpeeled for some time.

When I first saw this fractured oak post of a dismantled house frame, I thought that maybe a vehicle had rammed the side of the house. When a second post from the opposite side of the building showed the same type and location of fracture, I placed the two posts side by side for comparison. It became clear that the posts were sawn from the same log and the fracture occurred when the tree was felled. It probably hit a stump, another log, or a rock. These breaks are often difficult to see when the timber is green, so the carpenter might have missed it. Or he might have chosen to use it anyway. Either way, it has so far survived for close to 200 years.

Dendrochronology

A *tree's life history is revealed in its growth rings.*

A RELATIVELY RECENT ARRIVAL on the historic preservation scene is the use of *dendrochronology*, or tree ring dating, to date old wooden structures. This scientific and accurate method uses the annular growth ring patterns to fix the felling dates of the trees that the timbers, planks, and boards were cut from. In temperate forests, each year a tree puts on an early-wood and a latewood ring, in the earlier and later parts of the growing season, respectively. These rings can be distinguished by differences in color and cell structure, depending on the particular species.

Many of us have counted the rings on the stump of a freshly felled tree. The width of each growth ring correlates with the amount of rainfall in that year. A wet year has a wide growth ring; a dry year has a narrow one. A tree growing out in the open will grow faster with wider growth rings than one in the dense forest, but the proportional width of their rings will still follow the annual rainfall amounts. Growth patterns vary somewhat among species, and some, like black ash, are difficult to date because there is little variation in the width of the rings.

Using cores of standing old-growth trees and samples from a variety of old structures, a master chronology has been developed for common tree species used in construction in different regions around the world. While rainfall amounts may vary from town to town, a region's overall rainfall is relatively consistent (for instance, much of New York and New England can be grouped as one region). A sample of wood from an old structure can be matched against this master chronology to determine its exact age.

So, how does dendrochronology work? First, the specimen to be dated must contain bark edges, or wane, the wood just below the bark. It can be a timber, plank, or board, and should have 30 or more growth rings. Having more years of growth offers a better comparison and improves accuracy. The last growth ring under the bark will be the year the tree died or was cut down. If there is only earlywood present below the bark, the tree was cut during the spring/summer growing season. If there is latewood present, it was cut anywhere from fall to spring.

A hollow bit was used to bore into the waney corner of this post. Then a steel hook on a wire was used to extract the core.

A hollow-core drill is used to bore out and extract a small sample, perpendicular to the growth rings and less than the diameter of a pencil. The core is mounted into a slotted wood block and is sanded smooth, and perhaps even dyed to enhance the grain. Then, each ring is measured under magnification and recorded. A computer compares the patterns of the sample to the master chronology and hopefully finds a match to fix its felling date. Multiple samples are taken in each building, and if there are additions or reused timbers in it, those are sampled also.

A thorough dendrochronology investigation reveals much about a building. For instance, white pine timbers are often found with evidence of pine sawyer beetle larvae infestation. This indicates that the trees used were standing dead trees, or that the trees were cut and the logs sat on the ground for a season before they were squared up. The dendrochronology will bear out that these pieces are typically a year or two older than those worked from freshly felled trees. If the house or barn has been added onto over the years, the analysis will put dates to the various add-ons and modifications.

Tool Marks

AS SOMEONE WHO HAS WORKED with the traditional, historic tools of a carpenter for decades, I can gain a lot of information from the tool marks that I see in an old building. Was the carpenter left-handed or right-handed? Which side of the timber was he standing on? Was the piece cut quickly or labored over? Was it part of the original structure or added later? Though the carpenter is long gone, much evidence remains to answer our questions.

By examining the marks left by a broad axe on a hewn timber, for example, you can determine the handedness of the hewer, the height of the timber off the ground as it was worked, and the direction of hewing (forward or back) as well as the profile, length, and sharpness of the axe's cutting edge. By checking the marks on other faces, I can tell if timbers were worked on by individuals with different axes.

Axe Clues: Who Hewed This Log?

This illustration represents the hewing marks that might be seen on a typical hewn timber. The vertical marks are left by the scoring axe. The diagonal striations show the slicing marks of the broad axe. The end of the cut leaves the shape of the axe's cutting edge, it's "profile." Because of the arc of the swing, we know that the hewer was probably right-handed, working the log at knee height.

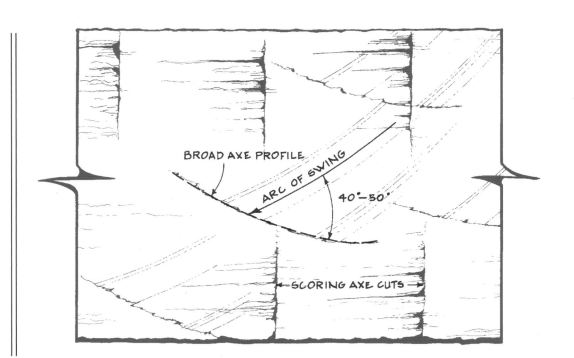

Axe Clues: Mortise and Tenon Work

The surface of this Dutch barn anchor beam tenon has the same surface texture as the timber itself, indicating that the tenon was hewn, too.

Sometimes you can even tell when the hewer stopped to sharpen his axe. Tiny nicks in the cutting edge will leave raised striations in the timber's surface. This becomes the axe's fingerprint of sorts and tells us the path of its stroke. By contrast, a perfectly sharpened edge leaves only microscopic striations. For new hewing work, I prefer my axe to be very sharp, without nicks in its edge. The work is easier and the resulting surface smoother. However, for detective work, I welcome the nicks!

Tool marks can also tell you how a joint was cut, from its roughing out down to the finished surface. Many timber joints show evidence that an axe was used first to remove the bulk of the wood. Of course: why would a carpenter pick away at it with a chisel when he could make quick work of it with an axe? Not only was the axe faster, it also saved that precious chisel edge from more frequent sharpening. An axe didn't need to be sharpened as often, and the job was done easily in a minute or so with a file. A chisel edge, being hardened carbon steel, required sharpening stones, careful technique, and many minutes to sharpen, and it had to be done more frequently. So it is not surprising to find faint axe cuts on tenon sides, mortise housings, joist ends, and scarf joints. The axe expedited the work.

Tool marks can even inform us of the extent of the carpenter's tool kit. One barn I examined had four different mortise and tenon sizes. While it might be typical to have two sizes, say 1½ inches for braces and 2 inches for everything else, this barn had 1½-, 1¾-, 2-, and 2¼-inch mortises. Remarkably, the plates alone contained mortises of three different sizes. There were principal post, stud, and brace mortises, each differentiated by their width. The only plausible explanation is that the builder had a set of those four chisels but no duplicate sizes. In order to keep multiple apprentices (perhaps four) chopping out mortises at once, he used multiple widths in a single assembly.

Puzzle Pieces

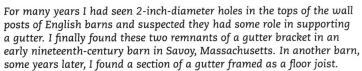

For many years I had seen 2-inch-diameter holes in the tops of the wall posts of English barns and suspected they had some role in supporting a gutter. I finally found these two remnants of a gutter bracket in an early nineteenth-century barn in Savoy, Massachusetts. In another barn, some years later, I found a section of a gutter framed as a floor joist.

For my own English barn, I reconstructed a gutter based on the old prototypes. Each time it rains, the animals' water trough is filled.

On another old frame, a small, simple Cape frame, the carpenter used only one size of bit for all the boring. Typically, a carpenter used two or perhaps more. The larger ones are for boring out mortises and typically match the width of the mortise (1½ or 2 inches). The smaller bits are for the pin holes and would be 1 inch or maybe ⅞ inch. In this house, both mortises and pin holes were bored out with a 1³⁄₁₆ bit. This unusual size is too big for normal pins and too small (and quite inconvenient) for even 1½-inch mortises. Undersized holes bored in the mortise means much more time and effort with the chisel.

The finish woodwork in the house was very well done, so I suspect one of the following two explanations: 1) The builder was primarily a joiner who was making his first stab at timber framing, perhaps working on his own house, or 2) the house carpentry and the finish work were completed by separate craftsmen. Perhaps the house carpenter was a former apprentice who had just become a master in his own right. The master under which he had trained would have been expected to furnish him with some basic tools to get him started in the trade. Because of its odd size, this boring bit was not highly valued by the master so he selfishly passed it on to his apprentice. It is fun to postulate on such things.

Adding and Subtracting: Understanding the Sequence

NOTHING IS EVER STATIC in nature nor in the built environment. Things are always in flux. We tend to think of old buildings remaining unchanged for decades or centuries, but the reality is that they are often modified as soon as they are finished or even as they are built. Why not? As buildings go from being small, two-dimensional versions on paper to three-dimensional, full-size structures, they are seen in a new light. The spaces are experienced in a way that could not have been appreciated before.

It is quite common for clients to make changes as the work progresses. Even when the same structure is repeated multiple times, as in a housing development, variations in the compass orientation of the houses makes each one unique. Orientation can dramatically change the daylighting in the

Sequencing Clues

Here, two joists tenon into a chase mortise. The opposite end is a plain (standard) mortise.

On the side of this second-floor timber we see a chase mortise. A stair header with tenons at each end was inserted between two beams after the building was erected. After one end was angled into its mortise, the other tenon swung through the angled trough until it popped into its mortise.

When adding on to a timber frame, it would be difficult to join a newly added plate to the original plate. Here, the addition plate is gunstocked, or jowled — made wider to accommodate more joinery. A tenon in the jowl enters a mortise in the post.

various rooms as well as the views from those rooms. So, even during construction, we may modify a design to better the final product.

Many times I have come across old timber carpentry that can only be explained by the clients changing their minds during the construction. Of course, the ever-obliging carpenter would do his best to minimize the effects of the owners' requested changes, patching and modifying things to make it work. Because of the way mortise-and-tenon joints must be assembled and pinned, they offer evidence of the sequence of the changes or modifications. For instance, some joints can be assembled only during the initial construction phase, so we know that unless the building was disassembled sometime during its life, the particular work in question is original. However, there are other timber joints, including mortise-and-tenon joints, that carpenters have used specifically to insert timbers into a construction after the fact. When we see these being used, the sequence is clear.

Another example of sequencing clues can be found in churches, meetinghouses, and other larger buildings that employ timber trusses. In these buildings, there is typically a ceiling attached to the bottom of the trusses. The relatively small ceiling joists that carry the ceiling finish must be flush with the bottom of the large tie beam of the truss. A simple mortise-and-tenon would seem to be the best joint to attach the joists to the tie beam. However, erection of the trusses becomes a critical issue. Trusses are best assembled flatwise and then tipped up into position. Inserting a multitude of ceiling joists into their respective mortises as the truss becomes vertical is a dangerous process, and pieces are likely to fall to the floor below.

By examining the joints, it is easy to see that the carpenters instead preferred to insert the ceiling joists *after* the trusses were vertical and properly seated. A number of special joints could be used, the most common being the *chase* or *pulley* mortise. Using this joint, the mortises on one side of the tie beam are lengthened to form a slot that would accommodate two or three joist tenons, allowing each of the three joists to be inserted at an angle and then swung tight into position (see photos on previous page). The opposite end of the joist fits into a normal-size mortise. The wood lath (for a plaster ceiling) nailed to the bottoms of the joists keeps the joists from swinging back out of the mortises.

MANY TIMES
I HAVE COME
ACROSS
OLD TIMBER
CARPENTRY
THAT CAN ONLY
BE EXPLAINED
BY THE CLIENTS
CHANGING
THEIR MINDS
DURING THE
CONSTRUCTION.

Reading a Carpenter's Mistakes

SINCE THERE ARE NO LONGER ANY eighteenth-century carpenters alive to tell us about how they worked, we must learn from the work itself. By studying the old stuff, especially when it is exposed during renovation or demolition, we can, over time, develop a good working knowledge of the carpenter's craft. Though mistakes might have frustrated the carpenter who made them, they often give us further clues about the historic building processes. I have seen evidence of mistakes in virtually every step of the construction process, from initial layout up to a finished joint, and these have been variously patched to hide them or otherwise fixed to make them function.

Oops! Cutting Errors

On this brace, you can see the layout of another tenon that someone had begun to cut on the wrong line. The line they should have cut is the adjacent, nearly parallel one. The piece was probably discarded and then picked up and reused.

The four circular cavities just to the right of the diagonal brace shown here indicate that someone started a mortise for the same brace in the wrong location. The circular cavities were chopped out with a gouge (curved chisel) to create a good seating for the shell auger bit. The carpenter realized his mistake before fully boring out the mortise. The lighter color of the area indicates that the layout scribe lines were planed away.

Oops! A Measuring Error

FILLER PIECES

"Measure twice, cut once," goes the old adage. In this simple half-lapped scarf joint, it looks like the builder made a one-foot error in calculating the length of the plate and had to insert some filler pieces.

Oops! Faulty Wood

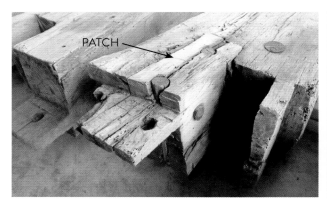

PATCH

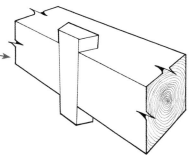

This seventeenth-century barn post contains a patch that appears to be original. In fact, it appears to have been done even before the tenon was cut. A rotten hollow in the core of this timber prompted the carpenter to patch it rather than replace it. Perhaps some of the joints on the post had already be scribed and cut before the rotten part was discovered. Replacing the whole timber would have been very time consuming.

Occasionally in old buildings I find a hardwood, under-squinted, tapered wedge slotted through the edge of a timber. This was a technique to take a bend out of a timber that had warped from seasoning. I would call this a "sliding dovetail beam straightener." Driving the wedge will lengthen that side of the timber, thus taking out the bend.

Oops! A Misjudgment of Space

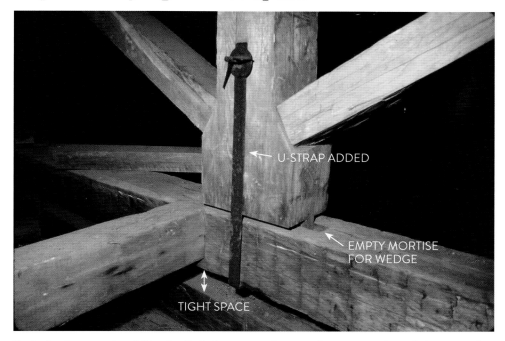

U-STRAP ADDED

EMPTY MORTISE FOR WEDGE

TIGHT SPACE

In the Congregational Church of Windsor, Massachusetts, the carpenter framed the trusses in an unusual way. Rather than notch out the lower chord of the truss (tie beam) for ceiling joists, which would weaken it, he used iron bolts to suspend a separate ceiling framing superstructure a few inches below the truss. Where the bottom of the kingpost meets the tie beam, he used the wedged dovetail through-mortise and tenon for its high tensile strength. In this joint, the tapered wedge, to be effective, must go in from below. The carpenter must have realized during the raising of the frame that there was not enough clearance to insert and drive these wedges in. Thus the mortise for the wedge just to the right of the kingpost is empty. A U-strap was added to hold the joint together, probably just after the raising.

Oops! Inconsistent Marking

Occasionally, scribe-rule buildings are numbered on the mortise and tenon faces to conceal the marks when the joint is assembled. Here, strangely, the carpenter used two different numbering styles to indicate a number four on each of the two pieces. While both are technically correct, the four single lines represent the standard numbering style for carpentry, while the IV would have been used for clock-face numerals.

How a Traditional Carpenter Thinks

This eighteenth-century illustration from Diderot shows many aspects of a French timber-framed building under construction, from preparing the components to raising the building.

TO BETTER UNDERSTAND the craft of timber framing, it would be helpful to get into the mind of a traditional carpenter. While no direct evidence of the inner workings of any historical carpenter's mind has been passed down to us, we have the evidence left in the buildings themselves, as well as the musings of this carpenter who has been studying old timber-framed buildings and creating new ones by hand for forty years. Here are some general rules a traditional carpenter would work by.

know when to fuss (and when not to)

Carpenters have always been practical workers. Rather than be fussy about all aspects of a job, they know when to work quickly and roughly and when to slow down and work more exactly. Typically, clients cannot pay for every part of the work to be exacting. While they appreciate beautiful work, it also needs to fit within their financial means. So a good carpenter knows how much tolerance is required for any particular job and takes only as much time as is necessary to produce the desired result. Much of the work may be hidden from view when it is assembled, and it would be considered extravagant to put unnecessary time and effort into those areas.

know your materials

A traditional carpenter is very familiar with his materials and knows the required tolerances for a given assembly. Is a quarter inch close enough, or a sixteenth? Working with green, unseasoned timber, for example, has different requirements than working cabinet mouldings of planed, seasoned wood. With green timbers, allowances must be made for the eventual effects of seasoning. How the wood is cut from the log will determine how it distorts as it seasons. Because shrinkage rates vary between species, so will distortion. Some surfaces must be hollowed out considerably to compensate for the eventual distortion. It isn't sloppy work — it shows that the carpenter understands his material.

A good carpenter can also "read" the grain of his material. He can tell from which direction he should run his plane to achieve the best results, which way to work knots to avoid tearing out the grain, and how to position joinery within the piece to maximize its effectiveness. He is knowledgeable of the different species of wood in his locale and understands which ones will be best suited to a given purpose.

This seasoned timber cutoff illustrates two effects of shrinkage on timbers: first, that a large check (a separation of wood across the growth rings) will often develop on the face that is nearest the pith; second, that the faces will often distort and no longer be square.

know your joints

When a traditional carpenter thinks of joining timbers, it is with wooden joinery: mortise-and-tenon, scarf joints, dovetails, notches, and so on. He has in his mind a catalog of joints that he has used or seen in his working career, although the catalog may differ somewhat from builder to builder and certainly varies regionally and among different cultural backgrounds. He knows the proportions of the joints (usually based on the framing square) and how they perform structurally. He knows where to use one pin, two pins, or no pins. He knows how best to cut and assemble the joint and how the eventual shrinkage of the timber will affect the joint's fit. After all, he has been using these joints for years.

be economical

Occasionally, some noncritical connections are simply butted and secured with a nail or two, but heavier-duty hardware, like brackets, straps, bolts, lags, and other iron fasteners, would not be the carpenter's first choice. These would require diverting some of his job income to another trade, that of the blacksmith. If he wanted to use some ironwork, he would have to set aside some time to visit with the smith and sketch out or describe the ironwork and arrange for delivery when it was completed. Because of the associated time and expense involved, ironwork, other than hinges or nails, was reserved only for essential applications such as trusses.

The same economic criteria still apply in today's timber-framing craft. Compare a mortise-and-tenoned beam-to-post joint to one using a bolted support bracket. The mortise-and-tenoned version could be cut on saw horses, on the ground, in about an hour; assembling the joint and pegging it

Hardware Versus Joinery

While much modern timber construction uses steel hardware to hold joints together, the result is often less than attractive.

Here, a similar situation is met with traditional wood joinery, a solution that is stronger, more attractive, and less expensive.

would take about five minutes more. Most framers would allow between $50 and $75 for the assembled joint. The steel bracket itself (with holes already bored and the piece painted) and the associated bolts, nuts, and washers will cost at least that amount and likely more. It also requires much more assembly time and working up in the air on ladders or scaffolds to bore holes for bolts, brace the assembly, align things, and tighten it up. In the end, the hardware version is likely to cost twice as much, be less strong, and certainly not be as attractive as the pegged mortise-and-tenon joint.

So why are hardware-connected timber joints used so much today? First, there is little understanding of traditional joinery by virtually all of those involved in building: architects, engineers, building officials, and contractors. Second, there is still a widespread notion that newer is better, especially newer technology. Third, hardware solutions are promoted in advertising in trade publications and textbooks. The companies that manufacture those products must advertise to remain in business. Meanwhile, the mortise-and-tenon isn't for sale in any catalog, so there is no vested interest to pay for its promotion!

This post base is anchored to the concrete with a bolted steel connection. While it may be effective structurally, it is not attractive or inexpensive.

The bottom of this black cherry post has been scribed to conform to the bedrock it bears on. A stainless-steel threaded rod has been grouted into the rock and runs up into the post. Inside the mortise, a nut and washer are tightened to prevent uplift. The mortise will be plugged with a cherry block to conceal the fastening.

DAVID E. LANOUE, INC.
ENGINEERS · ARCHITECTS · BUILDERS
Stockbridge, Massachusetts

Preservation and Restoration
of Traditional Architecture

(413) 298-4671

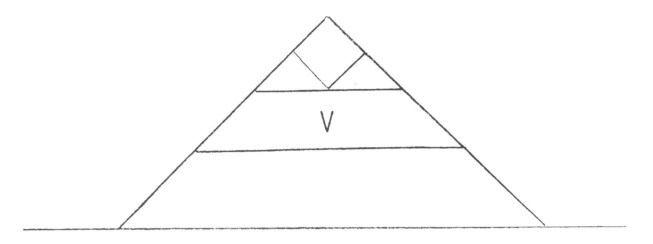

HAND TOOLS

HUMANS FIRST DEVELOPED primitive tools to make their hunter-gatherer existence easier and less taxing on their bodies. Simple tools allowed the making of spear points, the processing of meat and hides, and the grinding of grains. Though not the only species to make use of tools, humans have certainly perfected the practice, and over the eons tools have extended our reach far beyond mere survival: they have enabled us to create magnificent works of art and architecture.

The magnificent interior of the fourteenth-century Bradford-on-Avon tithe barn

Preindustrial Craftmanship

ABOVE *Norway boasts of 28 surviving medieval stave churches. The largest of these, Heddal stave church in Notodden, was likely constructed in the 1200s.*

WE WILL ACHIEVE EVER-LOFTIER HEIGHTS, some would argue, only by way of improvements in mechanization and technology. I tend to disagree. For while we can certainly see improvements in speed and accuracy of construction or improvements in safety or energy efficiency, the beauty and quality of building before the Industrial Revolution has not been matched. If I were asked to come up with a list of the 10 most magnificent wooden buildings in the world, some very impressive structures would come to mind: London's Westminster Hall and its hammer-beamed roof structure; the great temples of China and Japan; the stave churches of Norway; the huge tithe barns of Europe; the Church of the Transfiguration in Kizhi, Russia, with its many onion-shaped domes; and the incredible timbered roof structures

hidden above the ceilings of thirteenth-century Gothic cathedrals. Narrowing it down would be a daunting task indeed!

But rest assured, none of the possible choices for my top-ten list would be ones crafted by power tools or with high-tech gadgetry. Though we can build bigger, faster, and more elaborately today than ever before, we haven't matched the craftsmanship or artistry of preindustrial building. When one uses hand tools, it is nearly impossible not to be connected in a very personal way with the project's design and materials. When using power equipment, there tends to be a separation between craftsman and material. There is often less care taken, and less of one's self in the work. Of course, it doesn't have to be this way necessarily, but typically this is the case.

It seems that in 1937 England, people were already lamenting the fall-off in craftsmanship from that of earlier times:

> Each invention to make the task of the craftsman easier at the same time reduces the demand for skill; there is not now the necessity for the worker to be so highly trained. Many modern tools and appliances automatically perform tasks that once demanded concentrated ability.
>
> *Walter Rose, The Village Carpenter (1937)*

As a society, however, we cannot return to the past, at least not until we run out of the fossil fuels that power this great industrial machine. We can, however, slow down a bit to promote craftsmanship and artistry in wood, and we will be a better society for it.

Westminster Hall (commissioned 1393) in London is considered by many to be the finest timber-framed structure ever built. It is impressive because of its size (approximately 68 by 240 feet), the clear span of its hammer-beamed trusses, the degree of its finish, and the elegance of its Gothic design. Much of the timber work is moulded, and the horizontal hammer beams finish as angels.

An Extension of the Self

WHEN A PERSON USES A TOOL for an extended period of time, something happens to both the person and the tool. The person develops a knack for using that tool, an understanding of how best to wield it to create the intended result. Subtleties of grasp, body positioning, and movement develop and improve gradually over time. This coordination of eye, brain, and body is somehow stored in us for future similar tasks, so when that same tool is picked up again, the body remembers how to accommodate it. The longer a tool is used, the better the body remembers it.

After years of holding and using a tool, it seems that "the tool knows what to do," though it is actually the body. This can become evident when we pick up a different version of the same tool, one with a different handle or a different proportion. It can feel quite awkward and will take some time to adjust to the differences before the tool can perform to our liking. Some tools that are poorly made or badly proportioned will never feel right in our hands. They cannot perform their functions satisfactorily. They end up stuffed in the back of a drawer or rusting away in a heap. In my earlier years, I was attracted to such old tools, for they were often in nearly perfect condition, not having been used much. Today, I can spot them easily and avoid them altogether.

When a tool is used by a person, it becomes modified by that use and takes on some of the personality of the user. How it is held in the hands creates a patina on its surface. Metal surfaces become polished over time. Wood surfaces take on oils from our skin and become richly polished as well. If there are sharp edges where our hands make contact, we must smooth them over, lest our skin be constantly abraded. Over time we make subtle adjustments to the tool, improve its handle, change the angle of its bevel, and wear it down. In effect, it conforms to us and becomes a part of us.

Over a lifetime I have owned and used countless tools, but I can recall only two partings. I had used both tools happily for at least a decade, and they were well imprinted in my body's memory. After 30 years I can still picture them: a 2-inch chisel and a 28-inch crosscut saw. One was inadvertently left behind on a job, the other might have been stolen. Monetarily, they were only of modest value, but because we had spent much good time together they were priceless to me.

When you buy an old, used tool, it already has a history. Some are well worn, nearly worn out. They still tend to work well and are often easily acclimated to your hand. Others can show little or no use, perhaps just a coating of rust. This could be due to their awkwardness, low quality, or poor design, or it could be no fault of their own at all. Perhaps the tool just never got a chance to be used or tuned up by the craftsperson. After a few years of working in a craft, it becomes quite easy to look at a tool and figure out which is the case.

Though antique tools may be better understood and more highly valued today, this wasn't always the case. I remember a particular day in my childhood when I was fourteen. My grandfather on my mother's side (with the family name of Carpenter) had passed away, and two of my uncles were going through three old wooden carpenter's or joiner's chests in my grandfather's basement. An antique dealer had informed them that the old tools were of no value — he was interested only in the chests themselves. So I watched as my uncles filled a trash barrel with wooden planes, bit braces, and who knows what other treasures. I wish I knew then what I know now. One of my uncles did save a handful of tools to sell at a little Vermont country shop he owned. Years later, when he learned that I had developed an interest in old woodworking tools, he gave me a couple of those tools he still had from the earlier rescue.

> WHEN A TOOL IS USED BY A PERSON, IT BECOMES MODIFIED BY THAT USE AND TAKES ON SOME OF THE PERSONALITY OF THE USER.

Hand Tools: Character Sketches

HAVING SPENT OVER 40 YEARS using the traditional hand tools of a timber framer, I have certainly developed some opinions and preferences about them. Though I might be labeled a tool collector, with only a few exceptions all my tools have seen some use in my hands. My collecting, therefore, is more a by-product of my search for the more perfect tool than a desire to display stuff on a wall. During my career, I have learned much about the subtleties of handles, edges, balance, and other characteristics that make a tool perform well and a joy to use.

The Axe

From the Stone Age to our age, the axe has a long history. In simplicity and efficacy, it has no equal. A simple sharpened blade of stone, copper, bronze, iron, or steel attached to a wooden handle and swung in an arc is a very effective means of shaping stone or wood, butchering animals, defeating armies, or, most gruesomely, carrying out capital punishment. My own interest in the axe is, of course, for shaping wood for building purposes.

The axe tamed the forest and allowed mankind to utilize its bounty. Trees were felled by notching the trunk from each side, and then their branches were lopped off. Next, standing on the trunk, the woodsman cut V-shaped notches into the tree from opposite sides until it was severed in two. The pieces could then be hewn and used as framing members. The axe could also create simple joinery to connect the squared members.

The axe built the forts and palisades of North America, and it hewed the railroad ties crossing the land. Not too long ago, virtually every American boy was adept at using an axe by the time he was old enough to swing it. Chopping and splitting firewood was a daily chore.

In today's timber-framing craft, many are unaware of the axe's advantages in shaping joinery. Most of the axes available in stores are awkwardly made and subsequently used only briefly, with discontent, and then are set aside to rust. When I began using axes, there were still many older, usable, wood-handled versions lying about in barns, garages, and cellars. Hardware stores still sold some decent wood-handled axes. After all, it wasn't that long ago that wood was the primary fuel source here in the Northeast, so there was a regular demand for chopping and splitting tools.

Also, I remember old-timers telling me of how they earned a little extra income during the Great Depression. With their handy axe over their shoulder, they walked into the nearest woods, cut down a tree, and hewed out railroad ties. After carrying them out on their shoulders, they received a quarter per tie. These older axes had to feel right and work well — there was no alternative. And in those days, most people involved in making axes also used them and knew that a poorly proportioned or awkward axe would not sell.

SO WHAT CONSTITUTES A GOOD AXE FOR TREE FELLING, scoring a log, and roughing out joinery? What do we look for in a felling axe? First, the head should be made of mild steel that can be sharpened with a file. Hardened steel edges may stay sharp longer, but they can chip. They also require stones to sharpen them, which can take longer. Mild steel edges will dent but are easily and quickly sharpened with a file. The cutting edge should have a 30-degree bevel.

I prefer a single-bitted (single blade) axe with a poll to a double-bitted version. (The poll is the flat end of the axe, opposite the blade, and can be used for hammering; a double-bitted axe has two blade edges and no poll.) The head should weigh about 3½ pounds. Even the short-handled axes for roughing out joints should have the same weight. The thickness of the head right at the start of the cutting-edge bevel should be about ³⁄₁₆ of an inch; 2 inches in it should be about ½ inch thick. Newer axes are often too thin and tend to get stuck in the wood. Prying them back out after each stroke can waste time and energy. A thicker axe tends to pop off the wood chip rather than get stuck.

Most importantly, the cutting edge must be curved! I can't emphasize this enough. With a straight cutting edge, the whole edge strikes at the same time and produces more shock to the wood as well as the user. With a curve, the axe tends to penetrate farther with less effort and less shock. A sweep of about ½ inch seems best. If an axe poll has been used for pounding steel wedges in the past and shows some mushrooming, it should be beveled off with a file to prevent scarring the wood.

The handle of an axe is just as important as the head, if not more important. The overall length of the handle should be about 32 to 36 inches for felling, scoring, and splitting. People of height extremes might use a little more or less. For closer work, like chopping out joints, a 24- to 28-inch handle is appropriate. It should have a graceful curve with a nice swelling, or *fawn's foot*, at the end. The curve serves to make the hands almost precede the cutting edge in the swing. The heavier part of the tool, the head, will naturally trail the lighter part, the handle. This helps guide the axe in the proper trajectory (see drawing).

The fawn's foot also keeps the axe from sliding out of your hands and allows a lighter grip with less fatigue. A proper fawn's foot swells in both width and thickness, so the stock for the handle should be at least 1½ inches thick. Many newer handles are cut from 1-inch stock and don't feel right.

Geometry of a Good Axe

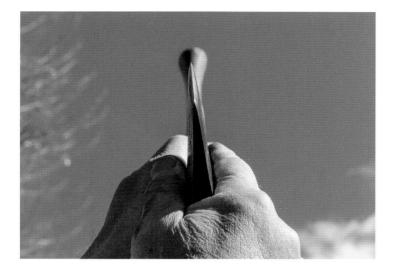

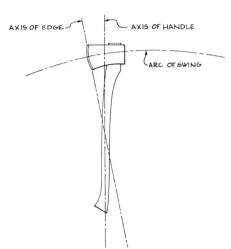

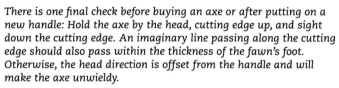

There is one final check before buying an axe or after putting on a new handle: Hold the axe by the head, cutting edge up, and sight down the cutting edge. An imaginary line passing along the cutting edge should also pass within the thickness of the fawn's foot. Otherwise, the head direction is offset from the handle and will make the axe unwieldy.

In a good axe, the combination of the curve in the handle and the angled axis of the cutting edge helps put the hands ahead of the cutting edge during the swing. The result is more accuracy and less fatigue.

Most importantly, the handle should be thinned considerably in the middle third of its length. This increases the handle's flexibility, thereby reducing the striking shock to the user. Though it may seem that thinning would weaken the handle, it actually makes it more resistant to breaking.

I like the handle to be about ⅝ of an inch thick by 1⅛ inches wide in the middle portion. A newer store-bought handle can be quickly thinned down with a spoke shave. If the handle is varnished or lacquered, you can remove the finish with sandpaper to lessen the chances of getting blisters. A penetrating oil finish is fine. The handle should be tight in the head and well wedged. Metal wedges have barbs to keep them in. If wood wedges are used to secure the head, it's best to apply a little wood glue before driving them in. It also helps if the handle protrudes a little past the head. You never want to lose an axe head.

There is one final check before buying an axe or after putting on a new handle: Hold the axe by the head, cutting edge up, and sight down the cutting edge. An imaginary line passing along the cutting edge should also pass within the thickness of the fawn's foot. Otherwise, the head direction is offset from the handle and will make the axe unwieldy.

Once an axe is tuned up and sharpened, you want to be selective about letting others use it, lest it get beat up by careless users. It's also a good idea to keep some spare axes for rougher work around the homestead so you won't be tempted to use your best one for splitting firewood or chopping roots.

A double-beveled axe that can be
used either right- or left-handed
with a 7-inch cutting edge

An American-made, eighteenth-
century, Germanic-style goose-
wing, right-handed axe

A late-nineteenth-century
Philadelphia-made axe with
a straight 12-inch-long blade

the broad axe

This variant of the axe is used to square up timbers, though a felling axe could also be used in a pinch. However, the broad axe differs in some respects. The blade is typically wider (hence "broad" axe) and the handle shorter. It also has a cutting edge of hardened steel because it is used more in a slicing fashion than a chopping one.

A broad axe can be a single-bevel or double-bevel type. If a single-bevel, it is sharpened on only one face; depending on which way the handle is inserted, it becomes a right- or a left-handed tool. Also, the handle should be offset from the plane of the cutting edge to prevent skinning of the knuckles while hewing close to the timber. This is accomplished with either a curved or bent handle combined with a head that has a symmetric eye (handle socket) or with a straight handle in a head with the eye twisted slightly. Just in case you were wondering, a *right-handed* axe means you work the *right side* of a timber, and your *right* hand is the hand closest to the axe head.

Contrary to what I said in my first publication (*Timber Frame Construction*, 1984), the back side or non-beveled side should not be perfectly flat and the cutting edge not straight. If the back of the broad axe is laid on a flat surface, it should rock in all directions. This allows the axe to scoop out the wood without the corners digging in. It also allows the axe to angle back out if it is going too deep. A slightly convex back side such as this will leave a gentle, wavelike surface. A perfectly flat one will leave a rough, choppy one.

A double-bevel axe (not to be confused with double-bitted, which has two blades) is sharpened on both sides of the blade, the eye or handle is not offset, and it can be used either right- or left-handed. In order to cut, it has to be tilted away from the surface, which creates the necessary clearance

No Flat Surfaces on a Broad Axe

The blade should have good curve along its length. On my favorite axe it is ⅝ inch in 7 inches.

The back is not flat but should have a crown of about ⅛ inch across the cutting edge.

The cutting edge curves down slightly from the back about ¹⁄₁₆ inch.

to avoid knuckle scraping. When tilted away, the curved edge produces the scooping effect nicely. Both single bevel and double bevel can be utilized to great effect. The single bevel requires less gripping power as the axe tends to hang naturally at the right angle to cut. It can only be used either left- or right-handed, depending how the handle is inserted. The double-beveled broad axe can be used both right- and left-handed, but tilting it to the proper cutting angle requires a firmer grip that can be more tiring.

For general hewing work, I prefer a cutting edge about 7 inches long with a sweep between ½ and ¾ inches. Overall handle lengths of 19 to 22 inches are appropriate. Because broad axes are used more in a slicing rather than striking way, the handle does not require the thinning that the felling axe does to prevent the transfer of shock.

Bevels and Blades

A single-beveled axe is shown on the left, a double-beveled axe on the right.

The broad axe should be sharpened to a nearly surgical edge and kept in a leather or wood sheath. Sharpening is done with stones and can take some time, so its edge should be protected dearly.

the adze

Adzes from left: My favorite adze; a late-nineteenth-century carpenter's adze; a shipwright's adze with cutting lips and spiked poll; and a gutter adze.

Another one of the simplest and most ancient of tools, the adze (or adz) has been used in many woodworking cultures. It can simply be a cutting edge lashed to a handle of wood or bone. The cutting edge could be a stone, copper, bronze, or iron piece. There are adzes of all sizes, from small hand adzes for carving, up to the carpenter's adze with a 30-inch or longer handle. The edges can vary from a straight to a gouge profile or, as with the shipwright's adze, the sides may turn up.

Our focus is the carpenter's adze. This tool, made of forged iron with an inserted steel cutting edge, or of cast steel, traditionally had a multitude of uses in timber framing and general carpentry. Though it resembles a grub hoe, a crude digging tool, it can actually do some fine finish work. In timber framing, the adze was used in laying floors in the days before the thickness planer. Rough-sawn boards will vary in thickness by perhaps ¼ inch or so. For a flat floor, it is necessary to have boards of a uniform thickness. To achieve this, the top surface would be planed flat and even. Then the edges were planed straight and tongue-and-grooved, or just grooved for splines. Finally, rather than plane off the entire undersides of the boards, they would be reduced to a consistent thickness only over the joists. This was accomplished with the adze.

If a joist or floor beam needed a little trimming to allow a floor board to lie flat, that too was accomplished with the adze. Adzes were also used to

The adze handle widens to match the tapered eye of the head, which, though wedged tight, is easily removed for sharpening.

skive the ends of clapboards (exterior siding boards), creating a long bevel to help shed water. Joseph Moxon in his *Mechanick Exercises or the Doctrine of Handy-Works* (1703) describes this use:

> They lay their Stuff upon the Floor, and hold one end of it down with the Ball of the foot, if the Stuff be long enough; if not, with the ends of their Toes, and so hew it lightly away to their size, form, or both.

Another curious use of the adze was related to me many years ago by an old chef: resurfacing large, freestanding butcher blocks. A block would get hollowed and irregular with heavy use, so an adze man would periodically visit and, while standing on top of the block, bring it back into shape. Though I have never tried it out, I can see that this is probably a very expedient method.

My favorite use for the adze is in creating the sweeping, curved reductions on the ends of joists and rafters (see photo above). This makes for a very graceful shape and is the best way to avoid shear-stress concentrations (created by downward force) in a reduction. A joist end with a right-angled notch may be easy to cut but promotes a horizontal shear stress failure, a split along the grain at the notch. Angling the notch improves it structurally, but the feathered-out curve is even better. In one eighteenth-century barn I surveyed, the rafters on one wall had angled reductions, and the opposite wall had curved ones. I am thinking that there were two framers doing the

Skiving Clapboards

Here a clapboard is in the process of being "skived" with an adze to create an end lap. The ½ × 6-inch board is placed in a holding jig, secured by standing on it with the left foot. The adze is then used to work down the bevel diagonally.

Installed, the skived end creates a weather-tight connection. The nailing is just above the top edge of the clapboard below.

rafters, each having his preferences, or perhaps there was only one adze on the job, the other carpenter using only the axe.

The way the adze handle is fitted to the eye, or socket, of the head is somewhat unique. Unlike the eye of the felling axe, which is oval-shaped, flared out at each end, and in which the head is secured with wedges, the adze eye is rectangular and has a straight taper. The handle is thickest at the head and tapers to match the eye. It is inserted all the way through the eye

NOT A HEWING TOOL

Other authors have claimed the adze was used for hand-hewing timber or finishing hand-hewn timber. I bought into this idea and early on tried my hand at adzing timbers. Despite years of work and modifying adze heads, I could not reproduce the marks and the appearance of old timbers that were supposedly adzed. If these old surfaces are examined close up in the right light, one can see that they show evidence of a slicing cut rather than the straight-on swing of an adze. The broad axe, not the adze, was the tool used for hewing.

number of feet. If the initial setting is accurately done and the walking out follows a straight line, the final dimension will be surprisingly accurate. Thus was measuring accomplished in the days before the tape measure.

The compass can also be used to scribe components together, following the irregularities of one surface and drawing the corresponding profile on the mating surface. This technique is still practiced occasionally in carpentry today. In times of old, builders used compasses to bisect angles, create regular polygons, divide lengths into equal segments, and other arcane tasks. Because it had sharpened iron or steel points, it could also be used as a simple scratch awl. If opened to a 90-degree angle, it fits nicely in the hand for such a use.

the line

Most buildings are comprised of flat planes. A string line pulled taut on a horizontal surface will be perfectly straight. If the line is saturated with fine chalk dust (or powdered charcoal, ochre, or ink), it can be snapped on that surface to leave a straight line. This straight line can then define a straight cutting plane through the surface. If a chalk line is applied to a vertical surface, gravity can make the line sag a bit depending on how tightly it is stretched, how long the span, and how heavy the string is.

An ancient and incredibly useful tool, the chalk line can even be used to describe a flat cutting plane on a curved or undulating surface. The work is positioned so the desired cutting plane is vertical. This eliminates problems

A traditional carpenter's chalk line, above, and a Japanese ink line called a sumit subo, below. The silk line is drawn through a reservoir soaked with India ink and charcoal dust. The tip of the line is secured with a needle-pointed wooden pin.

A chalk line pulled tight and snapped vertically will always define a straight, vertical plane. A line snapped horizontally will sag, depending on length and tightness of the string.

Snapping a Line on a Curved Surface

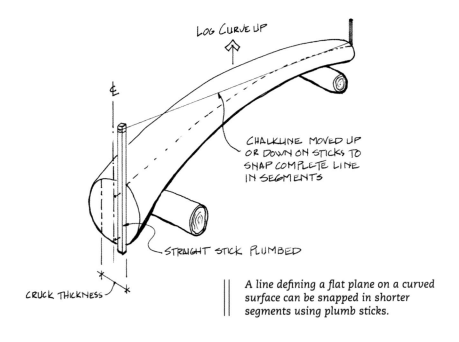

A line defining a flat plane on a curved surface can be snapped in shorter segments using plumb sticks.

with string sag. The line is pulled vertically to snap, and the chalk drops into hollows and follows irregularities in the surface. On dramatic curves, the line often needs to be snapped in segments.

Modern chalk lines are wound up inside of a small handheld case, or "chalk box," that serves as a refillable reservoir for the chalk and includes a crank handle that winds ups the line. The line is recoated with chalk every time it is wound into the box. The handle can be locked in place on the box so that you can suspend the box by the string to use the tool as a plumb bob. A simple chalk line can be made from a small, chalk-filled cotton bag that the line passes through. The ends of the bag are cinched around the line. Passing the bag along the taut string will chalk it up. White and blue chalks are water-soluble and are used to create temporary lines. Red and yellow chalks are waterproof and thus permanent.

the level and plumb

A spirit level and plumb from the late nineteenth century, above, and a modern version, below

The one commonality that all buildings on earth have is gravity. Gravity creates a vertical line of force that we can illustrate with a plumb line. For a given structure, these plumb lines are always parallel with each other. Though the curvature of the earth means they are not *perfectly* parallel, for normal-size structures they are effectively parallel. Lines perpendicular to

these plumb lines are *level*. In ancient times, a plumb mounted in a wooden frame was used to determine both plumb and level. It was a simple yet accurate tool that the carpenter made himself. Special wooden devices could be fashioned to give specific angles off plumb or level to gauge things like the batter (slope) of a wall or the pitch of a roof.

The spirit level, a seventeenth-century invention, uses a liquid, usually alcohol, inside a sealed glass vial. The vial also contains an air bubble. The vial is slightly arched so that the air bubble (being lighter than the liquid) always centers itself over the highest point of the arc. The vial has two lines marking a portion of its center. When the bubble rests in between the two lines, the tool is level. The vial can be mounted in a horizontal or vertical position to give readings for both level and plumb. Conventional spirit levels are still found in carpenters' and masons' toolboxes today, but there are also many new high-tech, high-priced gadgets available that use lasers and satellites. If the cost puts you off, remember that traditionally carpenters made their own level and plumb.

On the left is a simple wooden plumb made from a straight, parallel board with a scribed line down its center, with a lead plumb bob hanging. When the plumb bob string lies directly over the scribed line, it is plumb. On the right are two versions of a level: The top one is made from a single board with a line scribed perpendicular to the bottom edge. The lower version is four pieces joined but works the same.

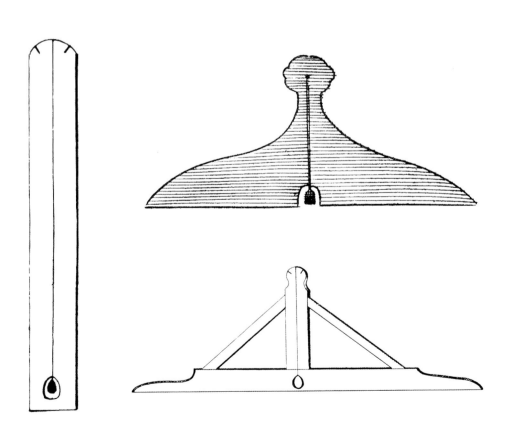

ONLY SAW THE LINES YOU CAN SEE

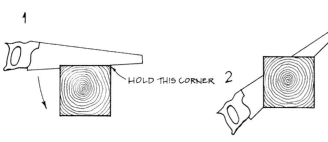

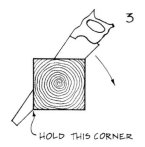
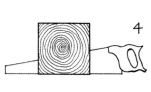

1. Start the cut on the edge farthest from you, working it along the line towards you and then down the front face, all the time keeping it slightly embedded in the kerf at the far edge.

2. Continue sawing down the front face, following the line you can see until you have sawed on a diagonal from corner to corner.

3. Walk around to the other side and put the saw back into the kerf. Saw down the face, following the line you can see, keeping the opposite side in the kerf near the bottom.

4. Continue pivoting the saw until you finish parallel with the bottom of the timber. No line is necessary on the bottom side. The finish cut will be planar and smooth.

handsaws

My favorite old crosscut handsaw was made by Simon Manufacturing of Fitchburg, Massachusetts, in the early 1900s and sold for $2.50. The price was stamped into the blade as well as marked upon a button in the handle.

In traditional timber framing, as well as general woodworking, there was quite a bit of sawing done. Though wood can be chopped, shaped, cleft, and pared with edge tools only, a carpenter will use a saw when it is deemed more expedient. Unlike riving, which follows the grain of the wood, sawing can cut across wood in any direction, regardless of grain, knots, or defects. It can cut exactly to a scribed line to create proper bearing shoulders for mortise-and-tenon joints, laps, or dovetails.

There is a knack to proper handsawing that few carpenters possess today. It would seem that a handsaw is so simple that anyone could use it without any prior instruction. However, that isn't the case. Even with a new, sharp, straight saw, an inexperienced person will mess up the cut. When instructing my students in proper use of a handsaw, I tell them to "saw only the lines you can see." While that seems like a no-brainer, virtually everyone sawing today violates that rule. They saw from one side only, not seeing the line on the opposite side of the timber. The saw needs careful guidance to follow a line. And the operator must be able to see the line! Therefore, after sawing diagonally halfway through the timber, one must switch positions to the opposite side to finish the cut.

There are different types of saws for cutting with the grain (called ripping) or against the grain (called crosscutting). *Crosscut saws* have teeth that are sharpened to knife-tip edges to score across the wood fibers. Because each tooth is alternately bent away to provide clearance for the blade thickness, there is actually a double row of cuts. The wood between them is carried away in the gullets (hollow spaces) between the teeth.

A *ripsaw* has teeth sharpened with a chisel edge. Because the saw is used with the grain, the teeth pare away the wood, rolling up the chips in the gullets. The ripsaw only works well in grain that is within 5 to 10 degrees of parallel. All other angled cuts should be made with a crosscut saw.

Since using a dull saw can be a grueling experience, good carpenters sharpen their saws carefully and often, and protect their edges when the tools are not in use.

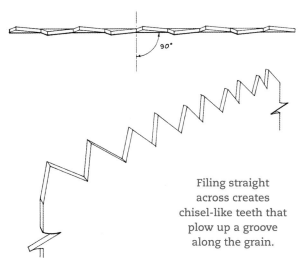

CROSSCUT SAW TEETH

CUTTING ACROSS THE GRAIN

65°

Because the tips of the teeth are filed diagonally across (here shown at 65°), they become alternating knife-like points that slice two parallel grooves in the wood. The wood between the cuts is carried away by the teeth.

RIPSAW TEETH

CUTTING WITH THE GRAIN

90°

Filing straight across creates chisel-like teeth that plow up a groove along the grain.

boring tools

Boring tools were used for making the holes for pins and for roughing out mortises (although I have found frames with no evidence of the use of boring tools at all). Before the advent of mass-production machinery, boring bits were time-consuming to fabricate and thus expensive. They were also a bit tricky to sharpen. The older type of bit used here in America was the *shell auger*.

Shell augers were forged by the local smithy and needed delicate filing to make them cut efficiently. Though they cut well enough, they also needed to be withdrawn from the hole occasionally to remove the chips. This was a slow and tedious process that may well have given rise to the other meaning of boring. Also, the auger bits lacked a screw point, so they couldn't start a hole. It was necessary to create a shallow depression, typically with a gouge, to keep the bit from wandering as it got started. While a straight chisel would do in a pinch to make a square or triangular depression, the gouge, with its curved edge, worked best. So if the carpenter had a shell auger boring bit, he probably also had a correspondingly sized gouge in his box. (The gouge could also be used to number the components of a frame.)

Another type of boring bit arrived on the scene about 1795 and was called the *Scotch pattern auger*. It was a great improvement over its

On the left is a shell auger; on the right, a Scotch pattern auger.

A TIME-TESTED MACHINE

My favorite boring machine is the Millers Falls boring machine made here in western Massachusetts. It features rugged construction of steel, cast iron, and wood parts and has a solid depth stop and a nice return mechanism that removes the chips from the hole as it withdraws the bit. It was listed in the Sears 1908 catalog for $6.65, along with the following claim: "This machine is conceded to be the best of its kind."

I purchased one at a garage sale in 1980 for $25 and, after a little repair and tuning, put it right to work. By a rough calculation, it has bored more than 50,000 holes for me. There have been no repairs required, just a periodic oiling. That is a lot of mileage for the money! I'll never know how many holes it bored *before* I picked it up, but it certainly doesn't owe me anything. How many modern power tools will still be in top shape after 40 years of use?

Also amazingly, I am still using the same two boring bits (a 1½-inch and a 2-inch) that I used from the start. I must admit, however, that I am very careful when working on old timbers (avoiding nails and grit), and the edges of my tools are always protected when not in use.

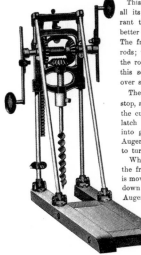

MILLERS FALLS CO. *p. 27 1886* 43

MILLERS FALLS BORING MACHINE.

This machine has been fully perfected in all its parts, and is now sold with full warrant that it will do better work and give better satisfaction than any other kind in use. The frame is made of half inch round steel rods; the braces are the same and attach to the rods at the top by a set screw. When this set screw is loosened, the frame falls over so as to bore at any desired angle.

The depth of hole to be bored is fixed by a stop, as seen on the left hand upright rod in the cut. When the frame strikes this stop a latch is lifted, and the machine throws itself into gear by the use of a spring, and the Auger is lifted out of the hole by continuing to turn the crank in the same direction.

When the Auger is drawn from the hole the frame hangs itself up until the machine is moved to the next hole, then it is dropped down by turning the crank back until the Auger strikes the wood, when it is thrown out of gear and proceeds to bore the next hole. As seen in the cut, the machine has adjustable cranks which fully regulate its speed and power.

The gears are all cut, which is not common in other machines.

We do not desire the reputation of making the cheapest goods in market, but we mean to deserve the name of making the best.

PRICES.

Machine, without Augers..... ....$7 50

Augers in sets...........18...........23...........41 quarters.
            $4 50 $5 50 $10 25

Sizes of Augers..(1, 1½, 2.)..(1, 1⅓, 1½, 2.)..(½, ⅝, ¾, ⅞, 1, 1¼, 1½, 1¾, 2.)

Boring machines, from left: A model circa 1840 with hand-forged parts (MA); a James Swan fixed upright machine (CT); a Millers Falls machine (MA); a Snell Mfg. angular model (MA); and the "Boss," a double eagle-model from Buckeye Mfg. Co. (OH, IN), with two chucks geared at different speeds.

predecessor. It removed the chips as it went, and it could start its own hole using its screw point. Though it undoubtedly cost more than a shell auger, the Scotch pattern expedited the boring process substantially. The gouge was no longer required for pre-boring and could be left at home. Tool boxes are heavy items, and heavier iron tools tended to get weeded out if they weren't used for a time.

If you walk into a building and see numbers made with a gouge on the components, the builder probably used a shell auger, and it follows that the building likely dates to before 1795. This is not a hard-and-fast rule, but it has held well in my experience.

By the 1840s, another development made the boring process easier: the *boring machine* or beam borer. The operator sat on this apparatus and cranked with both hands to bore out holes quickly and accurately. The boring machine easily bored holes perpendicular to the surface, and some models had depth stops as well. These machines were produced into the twentieth century and, judging by the surviving machines and the number of patents applied for, there were over 115 different boring machines made. Early models were locally fabricated with wood and forged iron parts. One model used a leather strap assembly to withdraw the bit from the finished hole. Later models used more cast-iron parts, and many could be angled or folded up for travel or storage.

chisels

Timber-framing chisels are large, thick, heavy chisels, necessary for the demanding work of chopping out mortises and shaping tenons. Though they can be used for paring with just hand pressure, they are more often struck with a mallet. Wooden chisel handles lessen the shock to the hand and help absorb the blows of the mallet. The two most common chisel widths were the 1½-inch and the 2-inch, corresponding to the two most common mortise widths.

I have used quite a variety of chisels over the years, both antique and new, and I have to say that I prefer the old ones to the new. Old chisels will likely need tuning up: removing rust, sharpening, replacing handles or ferrules, and oiling. They will also need a leather sheath to protect the edge. One could spend the better part of a day getting a chisel ready for use. An advantage of a new chisel is that it can often be used almost right out of the box. However, it will still need a little honing, and it will likely have some rough areas that could be smoothed off.

After four decades of working with chisels, I have developed an opinion on what makes a good timber-framing chisel, whether antique or new, and feel obliged to share it:

PARTS OF A CHISEL

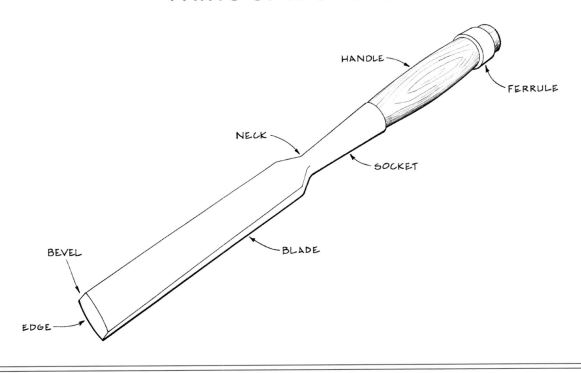

» The chisel width should match the mortise width. For instance, if you are making 2-inch-wide mortises, the chisel should be 2 inches wide. It is acceptable if the chisel is slightly undersize, but by no more than $\frac{1}{64}$ inch. More than that requires a lot of extra time spent chopping out mortises. If you are buying an old chisel, be aware that some may be oddly sized. If the chisel is grossly undersized, it should be reduced to the next smaller chisel size. If it is slightly oversized, it can be reduced to exact size. A machine shop is set up to do these operations accurately.

» The sides should be parallel, not tapered. No more than a $\frac{1}{32}$-inch taper, from the cutting edge to the handle, is acceptable. This will help guide the chisel in creating a parallel-sided mortise.

» The blade should be about 8 inches long before tapering gracefully to the socket. It should be about $\frac{1}{4}$ inch thick at the tip to about $\frac{5}{8}$ inch thick at the start of the socket. This extra thickness at the socket makes the chisel stand up to the prying and levering needed to clean out mortises.

» The handle should stick out from the socket about 8 or 9 inches, providing a comfortable grip for two-handed paring work. The longer handle moves the tool's overall balance point from the blade to the socket transition.

» When holding the chisel horizontally for splitting off tenon waste, it should feel balanced in your hand (and since it is uncomfortable to hold the blade portion, that is not where the balance point should be). Most handles found on both old and new chisels are too short to have this kind of balance, so if you desire this balance you have to turn (on a lathe) a new custom handle.

» The back of the chisel blade should not be flat. I know this seems counterintuitive, but if it is slightly arching (convex) along its length, the user can control how much the chisel is cutting by rocking it slightly on that curved back. If dead flat, the chisel tends to undercut mortises and leave choppy surfaces on tenons and housings. It is amazing how much of a difference this makes. The arch should be subtle, however — less than 1/16 inch over the length of the blade. This arching also allows the socket and handle to clear the surface being pared.

» Many older chisels have a laminated blade. There is a harder carbon steel piece approximately 1/8 inch thick, flush with the back side of the chisel, providing a long-lasting cutting edge. The rest of the blade and socket are a mild steel for absorbing shock. The cutting edge should be beveled to 30 degrees, but it can be a couple degrees steeper if you work only hardwood or a couple degrees shallower if you work only softwood. I prefer a flat bevel rather than the concave one created by sharpening on a grinding wheel. It is both stronger and makes it easier to determine the actual bevel angle.

KEEPING YOUR EDGE

Once you have a razor-sharp edge on a tool, you want to keep it there as long as possible. The chisel should have a protective sheath, preferably of thick leather, covering the entire blade. When the chisel is set down temporarily, the blade should be set on the leather sheath to protect the edge from grit. A single nick in the edge, whether it's from careless handling, bumping against another tool, or even a grain of sand,

can take considerable time to remove. A good craftsman would never throw a sharp-edged tool into a box of miscellaneous tools.

Most chisel handles were of the socket variety (top) rather than the tang type (bottom) and were fitted with an iron ferrule to prevent the head from mushrooming over and splitting when hit.

» The back of the chisel near the cutting edge should have a mirror polish. When the bevel is sharpened to 30 degrees and then honed to a mirror polish a couple degrees steeper, the two mirrored surfaces' meeting creates a surgical edge. That is the secret of a really sharp edge: two mirror-polished surfaces meeting at the proper angle.

» Each end of the cutting edge should have a crisp corner. A nicked or rounded corner will cause problems in mortise ends because it makes the cutting edge narrower than the blade, and the chisel will subsequently bind in the mortise.

» The cutting edge should have a slight curve of 1/32 inch to prevent the corners from digging in.

When looking for antique chisels, the manufacturer's name stamped in the blade can be an indication of a good tool. Some names to look for are T.H. Witherby, Peck Stow & Wilcox (PS&W Co.), Underhill Edge Tool Co., Buck Bros., Humphreysville MFG CO., Douglas MFG CO, G.W. Bradley, and James Swan CO. Though there are undoubtedly other good makes of chisels out there, these are the ones I am familiar with and have enjoyed using.

The above criteria define my idea of the perfect chisel. It may take some time to find one, and even more time to tune it up, but it will be a pleasure to work with!

the slick

The slick is a larger version of the chisel, fitted with a much longer handle. It is pushed with the hands and occasionally the hip but is never struck with a mallet. The handle should have a ball- or mushroom-shaped end for pushing with the body. Blades of slicks are commonly 3 to 4 inches wide and 7 to 10 inches long. The tool's overall length is about 30 inches. In other respects, its details of blade curve, thickness, taper, and so on are the same as the chisel. Its sides need not be parallel, however, for it won't be used in the ends of mortises. This is a hefty tool, and its weight creates momentum that will aid in pushing it through wood with minimal shock to your body.

A slick is not a required tool since a chisel can be used instead. However, for larger mortises, tenons, scarf joints, or any other large paring job, a slick will remove wood quickly yet leave a lovely surface. Because of its large size and cost, the slick is often the most prized and coveted tool in the timber framer's tool kit.

planes

Some of my well-used planes, from the top: a Stanley low-angle "Sweetheart" block plane, a Stanley No. 10 rabbet plane, and a wooden 1¾-inch wide-skew rabbet plane.

Planes have been around for thousands of years. The ancient Egyptians and Romans used woodworking planes. It is a simple idea: affix a sharpened iron blade at an angle in a wood-block body to take an even shaving of wood off a surface.

There are specialty planes, hundreds of types. There are planes to get into corners (rabbet planes), planes for tongue-and-grooving the edges of boards, planes for making window sashes and muntins, and countless varieties of moulding planes. Moulding planes allowed the carpenter to create wood mouldings for cornices and trim that would dress up buildings, inside and out. In earlier periods, the edges of timbers could be dressed up with moulding planes, as, in those times, the frame was the finish.

By the mid–eighteenth century in America, timbers were mostly hidden and cornices and trim were built up of boards and mouldings. Configurations of cornices changed as the building styles changed over time from Georgian to Federal to Greek Revival. Because a carpenter would fashion his own planes, they differed among carpenters and also varied by location. By examining and comparing moulding profiles and cornices on different buildings in the same vicinity, you might be able to identify houses built by the same carpenter.

Traditionally, carpenters and woodworkers of all types fashioned their own planes. Though the cutting edge was iron, the rest of the plane was wood. Beech was the favorite species to use because it doesn't seem to wear with use but rather gets polished to a sheen. As the industrial age approached, planes could be made with machines. At first they still had wooden bodies, then metal parts combined with a wood sole, and, finally, mostly metal with wood handles. Today, on all but the best planes, even the wood handles have been replaced by plastic. Also, most mouldings of today are machine-produced and there is a limited variety of shapes and contours to choose from at the local builder's supply. If the moulding profile you need is not available as a stock profile, a good machine shop can make a special cutter to produce it.

Surprisingly, the quality of a hand-planed moulding still cannot be matched by a machine. A rotating cutter, no matter how sharp it is or how fast it turns, will leave a slightly undulating surface. With a wooden moulding plane, the beech sole will actually polish the wood to a shine, leaving it perfectly uniform.

the spoke shave

A spoke shave is effectively a plane with a very short sole and winged handles. The short sole allows it to follow the contours of wood handles, chair legs, and other furniture parts. It is also very practical for smoothing up curved timber surfaces. A spoke shave can be made of wood or iron and can either be pushed away from you or pulled toward you. Some spoke shaves have convex or concave curved soles made to hollow out items like chair seats or to shape wooden handles. If you are a serious woodworker and haven't used a spoke shave, you are missing out big-time.

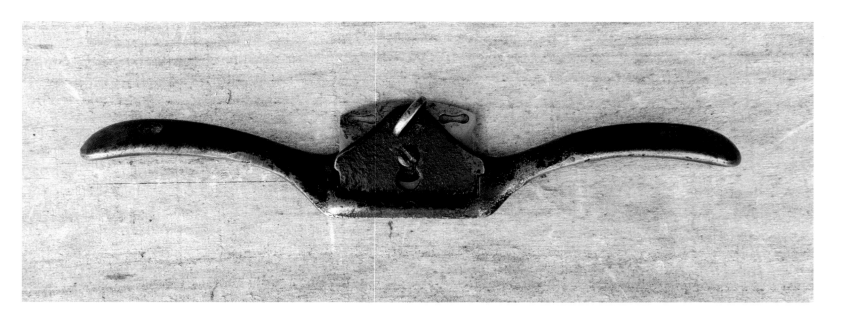

As with my smoothing plane, I have three extra spoke shave irons to allow me to put in a fresh one without taking the time to sharpen. When the last one is dull, I will take the time to sharpen the lot.

the drawknife

This wide-bladed tool has a handle at each end for two-handed use. It was used to remove bark from waney timber edges and create chamfers on those edges, and to shape pins and tool handles. For shaping, it is used in conjunction with the shaving horse, a foot-actuated clamping device that frees up both hands to use the drawknife. The cutting edge of the drawknife should be hardened steel and is therefore sharpened with stones rather than a file.

Pins are easily fashioned on a shaving horse with a drawknife. The pivoting "dumb head" is a clamp to hold the pin while it is being worked. The pressure is provided by the foot, which frees up both hands to use the drawknife. The drawknife is pulled toward oneself, removing much material quickly. It often takes less than a minute to shape a pin. The variously sized holes in the bench are used to test the completed pins. The pointed and tapered pin should go easily into the hole one-third its length.

hook pins

Hook pins or drift pins are typically made of iron or steel and were used to draw joints together temporarily, either during the scribing setups or during the setup of the frame. A wooden pin can swell up with moisture overnight and be difficult to remove. An iron pin can be removed by tapping it out from below, tapping or prying under the hook, or, if it is fitted with a hole, twisting it back and forth while pulling on it. In his 1703 text *Mechanick Exercises: Or the Doctrine of Handy-Works*, Joseph Moxon describes using hook pins during the assembly of the sills upon the foundation:

> They do not at first pin their Tennants into their mortesses with wooden Pins, lest they should lie out of square, or any other intended Position : But laying a Block, or some other piece of Timber, under the corner of the Frame-work to bear it hollow off the Foundation, or whatever else it lies upon, they drive *Hook-pins* into the four *Augre-holes* in the corners of the Ground-plates, and one by one fit the Plates either to a Square, or any other intended Position : And when it is so fitted, they draw out their *Hook-pins*, and drive in the wooden Pins and taking away the wooden Blocks, one by one from under the corners of the Frame, they let it fall into its place.

If working by the scribe rule, a carpenter would need a dozen or so hook pins sized to easily slide into pin holes.

These are some hook pins I had made by blacksmiths over the years. The two with holes allow another pin to be inserted to twist them back out. The two in the center must be tapped or pried out.

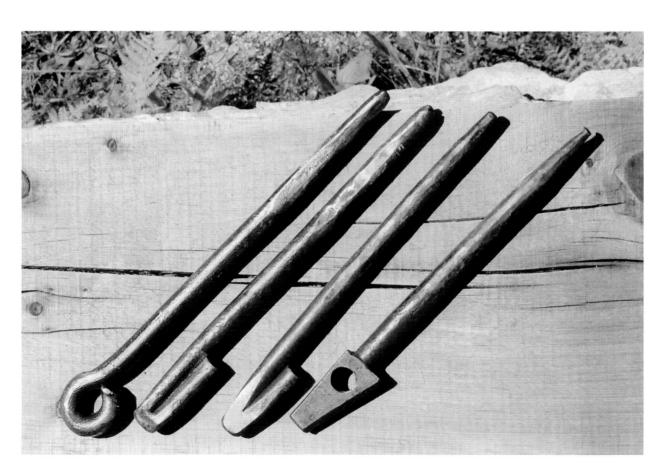

mallets, beetles, persuaders, and commanders

A "commander" dwarfs the regular mallets that are used to hit the head of a chisel.

These are all names for the large, two-handed wooden mallets used to drive joints together during assembly of the frame or to square up the floor frame on the foundation. A wooden mallet is less prone to damage the timber's surface than a metal one. When a metal mallet was used, it left tell-tale indentations.

A suitable commander can be fashioned quickly from a wooden replacement shovel handle and a rectangular timber cutoff, preferably from a hardwood like elm or another species that is resistant to splitting. The head is rectangular, rather than barrel-shaped, so the tool can stand upright and not roll. The striking edges of the head should be heavily chamfered, or "eased" with a bevel. The shovel handle is tapered in thickness and is inserted into a 1½-inch-diameter hole bored in the head, similar to an adze handle. If matched properly to the hole, the handle will only tighten as the tool is used. When the head needs replacing, you simply make another one from a new chunk of timber.

This 26-foot-long, light-weight spruce gin pole has been used by the author to raise a number of small buildings. It is positioned over the item to be lifted, leaning, and secured by guylines. The block and tackle is secured to at least three loops of a line passed many times around the pole. The extra turns around the pole prevent it from sliding down.

the gin pole

I was introduced to this simple hoisting apparatus by Richard Babcock (my former employer; see chapter 1), who was using it back in the sixties and seventies to raise up timber-framed barns without a crane. It is simply a straight, sound wooden pole, held nearly vertical by at least three guylines fixed to its top, from which a block and tackle is suspended. Leaning off vertical by about 10 degrees, the top of the pole is situated directly over the item to be lifted. The pole should be substantially longer than the height to which you are lifting, to allow for the lean of the pole, the drape of the tackles, and whatever straps or ropes suspend the load.

The pole should be a straight, knot-free specimen (or with only small knots) cut from the lower trunk of a small tree. I prefer spruce or balsam fir because they grow arrow-straight, are light yet strong, and often have only tiny branches. I leave a couple branch nubs near the top to keep the lashings from sliding down, and I allow the pole to season before I use it. A seasoned pole is much less flexible under a load than a green pole is.

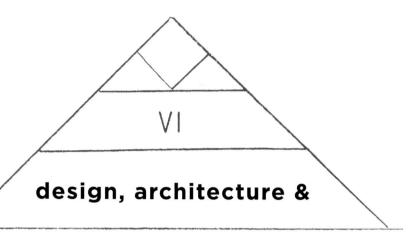

design, architecture &

GEOMETRY

All buildings are not created equal. There are those that have qualities, sometimes indefinable, that separate them from the others. We are curiously drawn to them. These buildings become an integral part of our lives, enhancing our existence. We feel good in and around them. What distinguishes these buildings from the countless other humdrum structures is design. It may be simple, straightforward design by local builders using a community's historical tradition, or it may be design by experienced professionals with years of training under their belts. In any case, good design creates buildings that are useful, attractive, and likely to last a very long time; and a well-designed home is something people naturally value and take care of.

What Is Vernacular Architecture?

Roros, Norway, is a town that grew up around a copper smelting operation in the seventeenth century. It is now a World Heritage Site and a very picturesque collection of buildings.

WHILE THE TERM "VERNACULAR" is often used to denote a local dialect, it can mean local *anything*. I am quite fond of vernacular architecture. It is a building tradition that uses locally sourced materials, takes into account the local microclimate, caters to the local lifestyle and customs, and builds with local hands. "Local" may refer to a region encompassing thousands of square miles or a hamlet of a few hundred acres. It is governed by landform and culture. Its boundaries may be distinct or nebulous, but you will know it when you see it.

Vernacular settlements are typically the most picturesque and quaint places you will find. They are the places you seek out when you travel. Examples might be a fishing village along a Norwegian fjord, an Italian hill town, a thatched-roofed English village, or an Arizona pueblo. These builders

used native materials, like earth, stone, timber, bamboo, and reed, without too much processing. The builders lived in the community and were often the buildings' owners. Buildings were tailored to the weather, designed to be warm in winter and cool in summer. Construction was local, simple, and efficient — it had to be. There was no hydraulic construction equipment or interstate transportation system, and no utility companies or banks. You couldn't make up for a building's shortcomings by dumping in more resources or applying technology. It wasn't practical or even an available option to them.

It was the Industrial Revolution that changed the world from local to global. The process has been slower in some areas than others, but the majority of our built environment today is no longer vernacular. A house in Arizona may be identical in appearance to one outside the District of Columbia. Materials are easily shipped nationwide and involve so much processing that their place of origin is indiscernible. In fact, many websites don't even list their company's physical address. Roadside architecture is even more ambiguous. A chain restaurant looks the same whether it is in Florida or Montana. Building codes, which started out local, are now, for the most part, national.

Of course, it's easy to "go with the flow" and build this lifeless, meaningless stuff, and that is what most people are doing. But I don't think that's what most of us really desire. Judging by the number of tourists who visit quaint vernacular areas (and the costs associated with visiting), it seems clear that this is what we really long for.

I am not advocating a return to the eighteenth century, as there are some great advantages to living in today's world. But along with the Industrial Revolution there came a sentiment that all the old ways were bad, and therefore much was thrown out that should have been kept. It was a mad rush, and indeed it is still a mad rush to embrace new technologies. I maintain that there are some wonderful qualities in old vernacular architecture that we are missing today. With a little bit of foresight, we can incorporate some of those qualities into our homes. The bones of a house can once again

Vernacular Architecture around the World

|| *Colorful wooden houses along a Netherland waterway*

|| *A winding street in Culross, Scotland*

|| *Densely clustered houses and windmills, Santorini Island, Greece*

|| *Ancestral wooden houses in Lemo, in the Taraja region of Sulawesi, Indonesia*

|| *Boathouses in the village of Solvorn, on a branch of the Sogne Fjord in Norway*

*Maramures Old Village,
Transylvania, Romania* ||

be shaped from the local forest trees; native stone can dress up the landscape and adorn foundations and fireplaces; and real, solid wood can fit out our house interiors.

What can you do? Search out the surviving vernacular in your area, find the early houses, and look at the local influences. What were the traditionally used materials? Since our country has been in constant flux since its inception, the local vernacular architecture likely underwent changes over time. There may be a Native American influence or a Spanish or French feel, or perhaps the place was settled during the 1840s and there is a preponderance of Greek Revival architecture. Some areas may never have had any permanent architecture or a defining building style, freeing you to create your own new vernacular there. If you do, you'll want to take advantage of local materials and the microclimate. Let's bring back vernacular architecture and make our homes and communities more desirable places to live.

Ancient Proportions

IN MY ARCHITECTURE HISTORY classes in college, we learned about famous Egyptian, Grecian, Roman, and Gothic monumental buildings and how much ancient architecture was based on geometry and proportion. Some of my fellow students took a keen interest in ancient proportioning systems and incorporated them into their designs. Others wanted to create something entirely new that they could call their own.

Builders of the eighteenth and nineteenth centuries were using these ancient classical proportions in their work. A lot of builder's pattern books taught carpenters to proportion their buildings according to an ancient system that was thousands of years old. Every book seemed to start with a treatise on geometry and the properties of two-dimensional shapes and three-dimensional solids.

Elegant Geometry

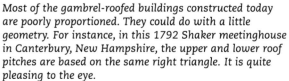

Most of the gambrel-roofed buildings constructed today are poorly proportioned. They could do with a little geometry. For instance, in this 1792 Shaker meetinghouse in Canterbury, New Hampshire, the upper and lower roof pitches are based on the same right triangle. It is quite pleasing to the eye.

In the study of old buildings, one often finds this geometric design scratched into a timber or even into stone. Done with a compass, it looks like a six-petaled flower within a circle. While this might appear to be the doodle of a child, it has far deeper significance. In England it is referred to as a daisy wheel, but in broader circles it is acknowledged as one of the two oldest and most widely distributed symbols: the sun symbol (the other is the tree of life). I believe its purpose here is as an orientation symbol. When I have found it, it has always appeared on a south-facing surface, unless the building has been moved. I think of it like the north arrow on a map or building plan.

THE 3-4-5 TRIANGLE

This is the classic graphic illustration of the 3-4-5 triangle and how it works
with the Pythagorean Theorem. The sum of the squares of the two sides of a right triangle will equal
the square of the hypotenuse. $3^2 + 4^2 = 5^2$ or $9 + 16 = 25$. For building purposes,
this particular right triangle is paramount.

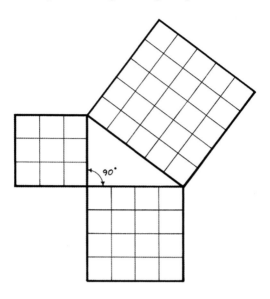

ROOF GEOMETRY AND PITCH

Roof pitch is the angle the roof makes with the horizontal, and it is expressed as rise over run.
The run is usually 12 inches, so the rise would be the number of vertical inches gained over
12 horizonal inches. Using a framing square, a carpenter can lay out exact angles for the rafter ends.
This example shows a 9/12 pitch, which happens to be a 3-4-5 proportion, a very common pitch.
When roofs were scribed, up until about 1800, pitch was expressed as the length of the rafter over the
span of the building. Rafter ends were scribed, so knowing all the angles wasn't necessary. However,
knowing the required length of a rafter for a given width of building was essential. So the same 9/12
pitch would be expressed in scribe rule terms as 5/8 — that is, 5 feet of rafter for 8 feet of width.

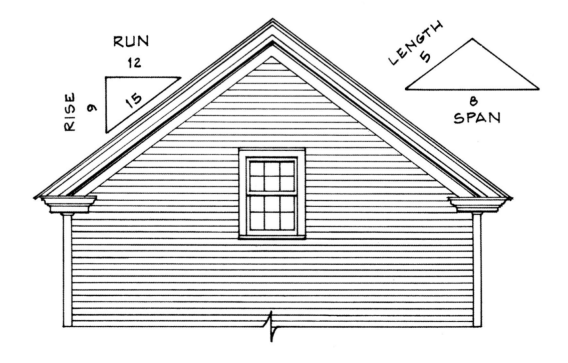

Why such an emphasis on mathematics? By applying these principles, and using a compass and straightedge, one could draft a design for a well-proportioned façade or detail a handsome cornice. Follow the ancient proportions, and your building was attractive. It was that simple. If you built without proportion, it was hit or miss, and most likely a miss. When studying old buildings, one expects the larger high-style and monumental buildings to be the ones with some underlying proportioning system, but, surprisingly, it was used even in the simplest of utilitarian buildings like barns.

pythagoras and the english barn

The English barn — a side-entrance, three-bay structure used throughout much of New England and New York from early settlement periods up to about the mid–nineteenth century — is a good example of a simple building with an underlying system of proportions. While it typically has no windows, no trim, and no cornice work, it is based on underlying proportions and thus, as a shape, it is pleasing to the eye.

So what is the geometry of an English barn? First, the plan (its footprint, when viewed from above) is a rectangle, and not just any rectangle, but one that has a ratio of three to four. If you make a rectangle with lengths of three feet and four feet, the diagonal will be exactly five feet. We learned that principle in high school (though it is probably taught in middle school today). Called the Pythagorean Theorem, it states that the sum of the squares of the sides of a right triangle is equal to the square of the hypotenuse of that triangle. The 3-4-5 is the lowest whole number set that works, but there are others, such as 5-12-13 and 8-15-17, that are less well known. In earlier times, the Pythagorean Theorem was not taught in schools and was known only to builders and the elite few.

Using the 3-4-5, or a multiple thereof, automatically creates a right angle, an essential element in building. The most common English barn size was 30 by 40 feet, and not by chance. The roof pitch on many of these barns was a 9/12 pitch. Modern roof pitches are denoted by a number of inches of *rise* (vertical distance) per 12 inches of *run* (horizontal distance). A roof that rises 9 inches for every 12 inches of run (a 9/12 pitch) has a 15-inch hypotenuse — it's that same 3-4-5 proportion. Some barns have horizontal wall girts spaced 3, 4, and 5 feet apart. I have seen one barn where the rafters are spaced out roughly 3 feet apart in the first bay, 4 feet in the next, and 5 feet in the third bay. So the master builder was having fun with the numbers, flaunting his power of proportion and knowledge of math and geometry.

In the best of houses, geometry and proportion appear in the details, too. This stair handrail terminates ever so gracefully in a decorative scroll.

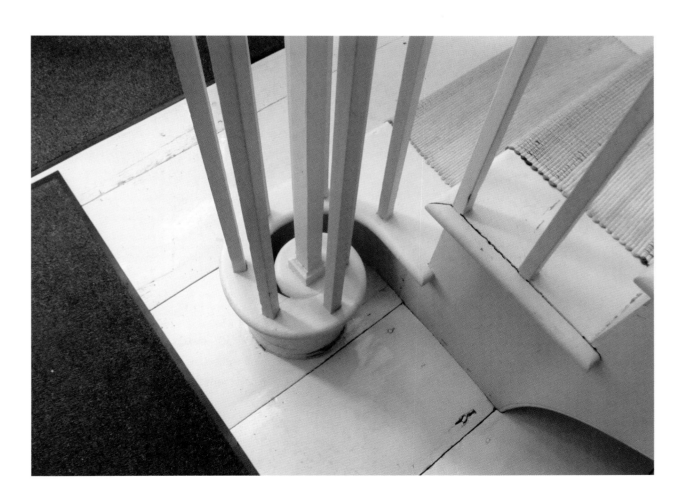

This drawing shows transverse and longitudinal sections of a late-eighteenth-century English barn in Adams, Massachusetts. The transverse section contains an equilateral triangle, as shown.

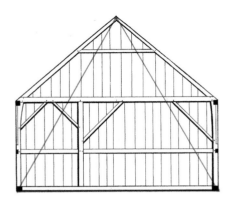
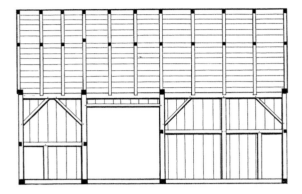

the equilateral triangle

The other underlying principle used in conjunction with the 3-4-5 on barns was the equilateral triangle. An equilateral triangle is one in which the three sides are equal, and the three angles are equal. In perhaps 90 percent of all the English barns I have measured and drawn up, an equilateral triangle inscribed in the building's transverse cross section will touch the very peak of the roof frame and the outside corners of the sill timbers. This geometry fixes the peak of the roof in relation to the building's width and is independent of the roof pitch.

The equilateral triangle has been used from ancient times — for the faces of pyramids, for Greek temples, and for Gothic arches. It even appears on the back of the dollar bill. Because it is not as high as it is wide, it creates a visibly stable shape. Not so tall that it might appear as if it could tip over, but not too squat, either; it was a perfect shape. Had I not applied my education in the history of classical architecture to my study of 200-year-old timber-framed barns, I might have missed this important discovery!

the golden proportion

If you want to incorporate geometric proportions into your designs, there are many to pick from. A very important proportion used extensively in historic architecture but also one that is the basis for much of what we see in the natural world is the *golden proportion* or *golden section*. This proportion is derived by taking a square, bisecting that square, then setting a compass to the diagonal of that half-square and swinging an arc. It creates a proportion roughly 1 to 1.618.

The ancients found this proportion in nature and chose to build with it. For instance, if you look at one of your fingers, you will notice that the spacing between the joints is not the same but rather a progression from the tip toward the palm. Each section increases by the golden proportion. Building designers used this proportion in both section and elevation drawings of their creations to great effect. Though good proportion may not be recognized consciously, the human mind subliminally recognizes its beauty and is pleased by it.

Another common early proportion is the Fibonacci series. The sequence is 1, 1, 2, 3, 5, 8, 13, 21, . . . etc., where each number is the sum of the two preceding numbers. The ratio between any two successive numbers approaches the golden proportion as the numbers get higher. For instance, 8 to 13 is equal to 1 to 1.625, and 13 to 21 is equal to 1 to 1.615.

I have noticed that when I find geometry and proportioning used in timber-frame structures, these proportions are based on the timber-framing dimensions rather than the measurements to the finished surfaces of siding, plaster, or flooring. A good clue is that the framing dimensions are often in whole feet rather than feet and inches, and measurements are almost always to the layout or reference face side of a timber (see page 103) rather than the center, as most architects and builders measure today.

THE GOLDEN PROPORTION, GRAPHICALLY

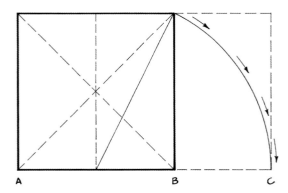

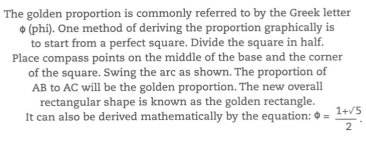

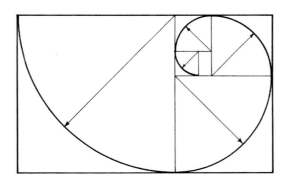

The golden proportion is commonly referred to by the Greek letter Φ (phi). One method of deriving the proportion graphically is to start from a perfect square. Divide the square in half. Place compass points on the middle of the base and the corner of the square. Swing the arc as shown. The proportion of AB to AC will be the golden proportion. The new overall rectangular shape is known as the golden rectangle. It can also be derived mathematically by the equation: $\Phi = \dfrac{1+\sqrt{5}}{2}$.

A golden proportion spiral is created by increasing or decreasing the radius of the arc by Φ for each quarter of a circle. It can be continued ad infinitum, larger or smaller.

Form, Space, and Light

GEOMETRY AND PROPORTION provide a good start, but myriad other factors affect the success of a building as well. Foremost among them are the building's overall shape and form, the principles of symmetry and balance, providing for natural light, and the outdoor space created by the building.

Inside, the floor plan is based upon a diagonal square set within the overall square, an ancient geometry. When the French doors are open, they recess into the walls.

mimicking mountain peaks

The best buildings mimic pleasing natural forms that we are familiar with. One frequently mimicked form is that of a mountain peak: it starts out low at the edges, gradually building and stepping up to a steep, snow-capped crescendo near the center. Many of the world's most iconic buildings echo mountain forms: the ancient pyramids of Egypt, Mayan temples, the great Gothic cathedrals of Europe, the stave churches of Norway, the mosques of Istanbul, Mont-Saint-Michel, ancient pagodas of Asia, the Empire State Building, the United States Capitol building, and, of course, castles (think of the Disney logo).

The mountain-mimicking Art Deco skyscrapers of the 1920s and '30s, with their stepped wedding-cake shapes, reached lofty heights. In fact, America's most photographed and beloved city skyline is that of lower Manhattan in the 1940s during the height of the Art Deco movement, when the skyline's overall shape resembled a mountain shape, with low, flatter buildings along the shore and tall, pointy towers toward the center. This quintessential skyline was familiar to all, especially the newly arriving immigrants who would never forget that image. Unfortunately, today's modern and boxy buildings at the water's edge have all but destroyed that iconic form of the city.

The mountain form, along with its smaller cousins, the sand dune and ocean wave, is applicable not only to large, monumental structures but also to modest buildings. A simple gable-roofed house with a center chimney alludes to this form.

MOUNTAINOUS ARCHITECTURE

The buildings below all mimic mountain forms.
Clockwise from top: Borgund stave church, Norway; Saint Basil's Cathedral, Moscow;
Mont-Saint-Michel, Normandy, France; and Disney's Cinderella Castle in the Magic Kingdom, USA.

Every building needs some sort of cap or pinnacle. A barn topped off with a well-proportioned cupola is more appealing than one without.

A barn with attached lean-to roofs of a lesser pitch and a cupola topping it off is another example. Have you ever noticed that a plain gabled roof without some sort of chimney or pinnacle is quite unsatisfying? The chimney caps off the house.

symmetry and balance

Symmetry and balance have been important considerations throughout the history of architecture. What occurs on one side of a central axis is mirrored on the other. The human form is symmetric, as are most animal forms. We see it in our fellow creatures, so we expect it in our architecture.

sunlight

Daylight is perhaps the most important and often most overlooked factor of good design. If we lived in a world where daylight came in evenly from all directions, designing for daylight wouldn't be such a big issue. However, the

The Illusion of Symmetry (and Size)

Often, historic buildings appear symmetric but in fact are not. A common trick in older New England Georgian- and Federal-style homes is to offset the center door slightly to one side. This created a false perspective because the human eye expects it to be symmetric. When approaching such a house from the side, as you would coming down the street, the house can appear either longer or shorter depending on which way the door is offset. As most buildings are approached from a common direction, say from the center of town, the center door would be offset to make the house appear larger and grander.

Balance

Balance is a form of symmetry where instead of a mirror image, there are forms of equal weight on either side of the center. Perhaps a low, fat shape is balanced by a tall slender one opposing it.

The façade of Lyndhurst, Tarrytown, NY exhibits the concept of balance. While there is symmetry within the individual components, the overall scheme appears random. However, the tall tower on the left is balanced by the shorter, slightly wider gable on the right, both equidistant from the center. The overall composition is pleasing.

The Taj Mahal in Agra, India, combines the mountain form, symmetry, geometry, and balance to a very pleasing effect.

Sunlight in the kitchen makes a wonderful setting. This arts-and-crafts-style home has southeast-facing windows over much of the counter.

sun rises in the east, makes its way in an arc across the southern sky through the day (or the northern sky, if you live in the Southern Hemisphere), and sets in the west. We are creatures drawn to light, so our actions tend to be governed by the sun. At least that is how it was meant to be.

Unfortunately, many of us work in cavernous structures with few, if any, windows, illuminated only by artificial light. Our bodies, however, have evolved over millennia to work with natural sunlight. Our cells require it to function properly. Artificial light contains only a small portion of the sun's spectrum, and being exposed to only artificial light can cause health problems. Therefore, we should be lighting our indoor spaces naturally wherever possible. In the best scenario, each room should have light on at least two

WE ARE CREATURES DRAWN TO LIGHT, SO OUR ACTIONS TEND TO BE GOVERNED BY THE SUN.

When light enters a room from windows on more than one wall, the overall light levels are more even and are much better for doing tasks. The window muntins (strips between panes) will further break up the light, reflecting it upward and sideways.

sides. This was more or less a standard feature in houses built before the advent of electric lighting but is not so common today.

A good house has bedrooms and kitchen facing east for the morning sun. Western sun is good for end-of-the-day relaxing in the living room or on a porch. The main entrance, garage openings, and outdoor work and play areas should be on the sunny side of the house, regardless of where the street is. If you place a square house out in the middle of a field, with a similar door on each face, I'll warrant that the doorway on the sunniest side is the one most used. Even if the driveway is on the north side, people will walk around to the south side to enter. We are drawn to light, so don't try to fight it.

Direct sunlight is a warm light composed of many colors (remember the prism experiment from science class). Light entering windows from the north side is indirect sunlight reflected from the sky and is heavy with blue light. While it is an even light preferred by artists for their studios, it can be depressing over time. It is not recommended for rooms that we live and work in.

Sun orientation is overlooked by most people. Most would be hard-pressed to say which direction the windows face in their houses. When they buy house plans to build with, they face the front of the house toward the road, regardless of the compass direction. As a result, the kitchen may be sunny and cheerful, or it could be dark and depressing. If you don't pay attention to the sun, it is a gamble, and judging by the current housing stock out there, most have lost on that gamble!

There is a landmark work that deals with how humans and buildings interrelate that is applicable to all cultures and locations around the world. *A Pattern Language: Towns, Buildings, Construction* (see References & Further Reading, page 258) deals with factors that have affected us since we began to build. This book has profoundly changed my thinking about some aspects of design and building. Written by multiple authors and spanning many cultures, it is essential reading for all who contemplate building.

pleasing outdoor space

A building with a single compact form is economical to build, heat, and maintain, but it doesn't define outdoor space very well. A building of two shapes joined to form an L does. It partially encloses a south-facing outdoor space, sheltering it from cold winter winds and trapping more sunlight. The space gets not only the direct sun but also the sunlight reflected off the surrounding walls. When sitting outdoors, people like protective enclosures to their backs and sides. It can be equated to sitting in a high-backed, stuffed wing chair facing the view. Outbuildings, fences, hedges, and other elements can provide that comforting enclosure while still allowing for views. This is one of the principles of *feng shui*, the Chinese art of placement.

RIGHT, TOP *The two parts of this New England house meet at a right angle to form a sunny, sheltered courtyard where chores can be done (Freeman Farm, Old Sturbridge Village, Massachusetts).*

RIGHT, BOTTOM *The sunny courtyard of this country cottage in the UK is enhanced by the restful pale-blue trim and white-washed wall that reflects the light. The low eaves and even lower stone wall create a comforting enclosed space that doesn't feel confined. It also offers a warm, sheltered place perfect for a garden.*

color

Building materials should enhance our quality of life and boost our spirits. Color can be important in this respect. There is quite a science about the effects of color on humans. Artists, advertisers, and graphic designers all use it to great effect. Color done right improves life; color done wrong worsens it. Many people don't get the connection! As a boy going to school in the sixties, I remember many classrooms painted the same ugly green color. Years later I learned that particular color, called "institutional green," was determined by government researchers to be the color best suited to keeping those wilder youngsters under control. That's something to keep in mind if you have one. Color is subliminal.

So what are the best colors? Earth tones with rich, warm hues, like those found in nature, often create the most comfortable environments. Unlike most man-made materials and colors, natural colors are not uniform nor pure. They are made up of several colors and can change their appearance depending on the sunlight or season. They are easy on the eye. Those colors are hard to duplicate in the factory. Materials taken from the earth (like stone) or harvested from the forest (like wood) come with a great variety of the natural colors automatically. It is hard to go wrong when using them.

good local materials

Good materials are locally sourced, with a minimum of processing, and can be harvested sustainably. They aren't a rare resource, they don't create toxins during their processing or harm humans that handle them, and they don't need a lot of wasteful packaging that ends up in landfills. Many so-called "green" building materials marketed today are really not that green. For instance, a natural, beautiful, nontoxic building product may have been shipped several thousand miles to your building site, perhaps crossing an ocean. No matter how they spin it, that just isn't green!

These good materials should last indefinitely or at least be easy to renew. If they wear easily, they should wear gracefully. Honest materials have their color throughout, not just a surface application. Over time, they take on a patina, which makes them even more handsome than when they were first installed. They may absorb the oils from thousands of hands rubbing across them (like a wooden handrail) or gain a scalloped depression after being stepped on for decades (like a stone stair).

Materials used on a house's exterior should weather well and gracefully. If properly applied, wood siding can last for centuries without any toxic chemical preservative. It will wear away over time from ultraviolet rays, windblown sand, frost, and microorganisms — roughly a quarter of an inch every century. It may get deeply furrowed, like driftwood, and its knots,

RIGHT, TOP (LEFT) *The bold, earthen colors are in harmony on this wattle-and-daubed timber-frame wall at the Norse Folk Museum in Oslo, Norway.*

RIGHT, TOP (RIGHT) *The interior of this house in Halsingland, Sweden, has very calming azure blue doors and trim.*

RIGHT, BOTTOM *The buildings of Venice have muted, earthen colors that make them appear softer in the warm Mediterranean light.*

The arched black cherry tie beam, the curved braces, and the well-crafted stonework make the entrance to the Mountain Top Arboretum education center very inviting.

being relatively hard, will wear less than the wood around them. Though the wood will become thinner, it should still be sound. If a board here or there gets cracked, worn through, or rots because the earth was too close, it can be replaced with another without too much trouble.

I still remember the first time I removed some original white pine siding boards from a barn built before 1800; it had never been painted or stained. The boards had beautiful color variations and were gracefully furrowed, highlighting the grain, but they were only about half an inch thick. Where they were tucked up under the eaves or behind a trim piece, out of the weather, they were still a full inch thick. Also, the knots were raised and highlighted. Each board was a natural work of art. I subsequently learned that boards such as those are valuable. Wealthy clients would pay hefty sums (10 to 20 times that of a newly sawn pine board) for those boards to adorn their country house interiors. Again, it would be hard to duplicate this look in the factory.

Do all materials in a home have to comply with the above ideals? Not necessarily, but as many as possible should, when practical. Unless you live in an industrial area, you would be unlikely to get your plumbing or electrical components locally, so there will be some compromises. But if you keep these concepts in mind as you select building components, you will be more likely to live in a house that will enhance your life and age beautifully and gracefully.

details and finish

When selecting finish materials for a home, it is important that there be some consistency in the level of finish on those materials. For instance, rough wood surfaces are fine as long as the adjacent material surfaces are equally rough and irregular. Rough timber, stone, brick, wrought iron, clay tile, leather, cast bronze, and unpainted plaster are all compatible. Combining polished, slick, and angular materials with rougher, imperfect surfaces can be challenging. And, generally, the more perfect the materials are, the fussier their detailing becomes. A rough end cut of a perfectly flat, uniform material is as incompatible as a fine, precise cut on a rough material. When we use uniform, modern materials, we expect them to be perfect. The eye easily picks up any irregularities, blemishes, or imperfections, and we find them disconcerting. The tradesperson must be ever vigilant in selecting and installing such materials. As an occupant, we must also be vigilant not to mar, crack, or stain such perfect materials.

If we use less-than-perfect materials, like plaster instead of drywall, slight undulations of the surface are not bothersome to our eyes and are, in fact, preferred. The axe-hewn surface of a timber can have an undulating surface, or it can be twisted from drying or have multiple seasoning checks, and no one seems bothered. Old wood surfaces can be furrowed from weather, worn from use, or riddled with holes from insect borers, and they are appreciated all the more.

Wood and Stone: Partners in Building

When one thinks of permanence in buildings, it is usually stone or perhaps masonry that comes to mind. Indeed, the oldest intact structures in the world are made of stone (think pyramids), and the majority of ancient ruins are likewise stone. If a building of stone and timber is left unmaintained and falls to ruin, the timber will decay and eventually disappear, leaving only the stonework. This stonework, left to the elements and subject to weather erosion, freeze-thaw cycles, or ground subsidence will degrade over time, albeit much more slowly. However, if a building combining stone and wood is cared for, it can last indefinitely. Roofs framed of wood can shelter stone walls, keep them dry, and prevent their degradation. Wooden portions can be repaired or replaced more easily than stone.

Each material has its advantages and disadvantages. Stone is hard and durable and resistant to decay. It is extremely strong in compression but relatively weak in tension or bending. It resists gravity loads well, but even the best stone buildings can easily topple in an earthquake. Stone will resist wear under the footsteps of countless humans, but its unyielding rigidity also makes it tiring to stand on for any length of time. It can be found all over the planet in a huge variety of colors, hardness, and workability. While readily accessible stones lying on the surface of the earth have always been used for building purposes that require minimal shaping, stones that need to be worked into specific shapes are more often quarried. As with timber, stone is also recyclable. From ancient times to modern, when an old stone building is no longer used, falls into disrepair, or collapses, its stones will be scrounged and used again and again.

Two buildings, one with wood walls and the other with stone, each have their place as an example of good vernacular architecture.

All the stone for this project came from the excavation of the foundation, so its colors and textures are in harmony with the environs.

Wood, in comparison, is strong and durable, but only when kept dry. If it is allowed to stay wet for extended periods, decay organisms and wood-burrowing insect larvae will attack it, reducing it to mush. It is relatively strong in compression, tension, and bending, yet it is flexible enough to withstand the forces of earthquakes. It is also found over much of the planet in a variety of densities, colors, and textures and with varying durability. It is softer and friendlier to humans than stone and insulates much better from the cold. But, most importantly, it is easier to harvest and transport and easier to shape with tools, and that makes it relatively inexpensive to build with compared to stone.

In the best of buildings, one would combine stone *and* timber, using each where it is best suited for appearance, strength, and durability. In finely constructed wooden buildings, freestanding posts have their bottoms bearing on a stone block. The block separates the wooden post from the dampness of the earth below to prevent decay. Though the stone can wick up a bit of moisture from the earth, it tends to wick much less than concrete does. The higher the stone portion rises above the floor, the less moisture will reach the post bottom.

In New England, stone is found almost everywhere. When we use it in our buildings and landscaping, our creations become part of and at peace with the natural landscape.

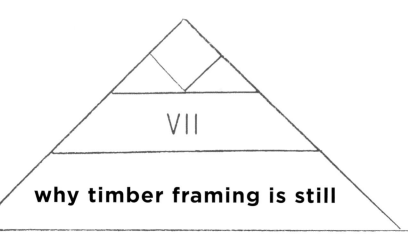

VII

why timber framing is still

RELEVANT TODAY

CERTAINLY THERE ARE FASTER, easier, and more efficient methods of construction than mortise-and-tenon timber framing. We now have technologies and machines that elevate us above, and virtually remove us from, the act of building. There are some pretty expensive, powerful tools and complicated jigs that can perform incredible tasks. Surely 3-D printing of whole buildings is on the horizon. And much of today's building work is done inside factories or, if done on-site, it utilizes components and assemblies made in factories. The building craftsman has become a prefabricator or installer. Timber framing, however, as well as other natural and traditional building forms, restores some of the tangible benefits of building that have been lost over the years. It harkens back to the days of romantic building methods and honest materials that we long for.

We feel more secure when we see the bones of a building supporting the roof above us.

Honesty in Building

IN MOST MODERN BUILDING SYSTEMS, the structure is concealed. The walls and ceilings are flat surfaces, and the structure is hidden behind them. The occupants haven't a clue as to the type of structure or its inherent strength. By inserting features such as concealed girders and trusses, designers can actually trick the mind with wide-open spaces, windows that wrap around corners, and the like. Roofs can seem to float, with no visible means of support.

Though many people spend their lives in such structures, they are typically happier when they can see the "bones" of a building. They can see a post where there should be a post, a beam where there should be a beam. They have some assurance that the structure is sound and more than adequate to protect them. No one's mind likes to be tricked! Exposed timber-framed buildings are about as honest as it gets. There is no mystery as to what is holding up the ceiling or the roof. We can see the diagonal bracing that gives a building rigidity in the wind or keeps it upright when an earth tremor shakes the ground beneath it.

Though today's houses are rarely built with usable attics, they can be wonderful spaces. Besides providing storage space, they can be play areas where children can imagine other worlds, or retreats where grownups can delve into their families' past while looking at heirlooms. They also provide a tempered space between the outside air above and the conditioned space below. This nineteenth-century attic space is beautiful in its own right.

Connection to the Natural World

WHEN USING MATERIALS to build with that were once alive, we are connected to the natural world. Timbers are squared trees with more than a subtle resemblance to their original form. No one will walk up to them and say, "What is that?" Wood is familiar to all, everywhere. It is also one of the greenest building materials. When cut from local forests and prepared with very little processing, (which might even occur outdoors, right on-site), wood has a very low *embodied energy*, the value of all of the resources used in creating, processing, and shipping a product. By contrast, materials such as concrete and steel have high embodied energy because their processing requires lots of fossil fuel and high-temperature production, and they must be transported by large, energy-consuming vehicles, sometimes from great distances.

Just choosing to use wood, however, does not guarantee that it is environmentally friendly. There are places on this earth where forests are not being managed properly or sustainably. As consumers, we should be ever vigilant to purchase all our products from sustainable sources.

As for its carbon footprint, wood manufacturing processes emit relatively small amounts of carbon, and wood is actually a carbon storehouse. Live, growing trees take in carbon dioxide and give off oxygen. And the carbon stored in them stays in the wood after the trees are cut. The forest then grows more trees and stores more carbon. If a timber-frame structure is dismantled, its components are easily recycled, so the carbon remains locked up in them. Only by burning or decomposing can the carbon be released. In short, when you build with locally sourced wood, you are helping the planet.

If trees are harvested on or near the building site, the timbers can be worked up on-site with minimal handling and no trucking. On one side of this portable sawmill is the log pile; on the other, the finished timbers and boards.

A weathered and forked black cherry post frames a view of a traditional nineteenth-century-style timber-framed barn, which is actually a two-car garage with a workshop on the right end.

The equipment shed for the Mountain Top Arboretum has an organic, rough-sawn timber frame made from on-site trees. The tall, tapered column at right has its bottom scribed to a fieldstone that is embedded in the concrete floor.

Connection to the Past

A "barn raising" was a chance for a community to work together to help a farmer and his family erect a new barn — then eat, drink, and celebrate together afterward. It is good to keep such traditions alive.

TIMBER FRAMING IS NOT JUST a type of construction. It is a tradition that goes back thousands of years. Our ancestors on every continent containing temperate forests were familiar with timber-framed buildings. Even in those areas where stone buildings predominated, roofs were often timber-framed. For thousands of years we have had a relationship with trees, using the wood to cook our food; warm and shelter our families; enclose our fields; and build our seagoing vessels, furniture, wagons, and barrels. Trees are also a direct food source for us and our animals. Since the beginning, we have lived off the acorns, nuts, and fruits of trees.

Though we seem to have more of a relationship with plastic these days, our ancient relationship with wood and trees is still in our genes. When we sit in front of a blazing, crackling fire, or run our fingers over a polished wood surface, we are stirred by that connection. I have noticed over the years that there are few people who are not moved when seeing their first timber-framed home. And indeed, many develop a fascination, if not an obsession, for the craft. We envision ourselves as rational and thinking beings choosing our own paths in life, but those paths are guided by our genetic predispositions — and wood figures predominately in our genes!

Over the years, the clients for whom I have designed and built timber-framed homes have been very appreciative. However, there was one older couple from the Midwest USA who were especially appreciative. Their house-frame raising party back in 1988 was quite an extravaganza. It was a catered affair with a large tent and live music. Over 200 guests attended and hailed from all over the country. The raising went smoothly and quickly, but the partying and dancing lasted long into the night. Two years later, we raised up a second building on their property, a two-car garage with a guest house above. At this second event, also a catered affair, my clients informed me that their house-raising was "the second most important day in their lives." Wow, I was taken aback with that comment. Of course, I had to ask what the most important day was. "The birth of our first child," they responded. I was honored to have been part of their second-best day!

I have always tried to involve my clients in the physical end of the work. It often starts in the forest, occasionally the clients' own woods, where they can see me and my crew shaping timbers with axes, or teams of oxen drawing out logs for their timbers. The process continues with the cutting of the mortises and tenons, where the clients see hardworking craftsmen wielding antique woodworking tools. Finally, it ends with the hand-raising of the completed frame. The client often assists in the tradition of nailing the evergreen bough to the peak of the frame at its conclusion. When finished, it is more than just a house. The owners have borne witness to the fact that we put part of our lives into crafting their home. When they look at a timber, they see the trees in the forest and the faces of the craftsmen who worked up those timbers. They have made so much more than just a financial investment in their home: it is a part of them.

So when a timber-framed home goes up, there is a lot more to it than just a physical structure. There is much more than just dollars and cents. Those involved are part of a continuous tradition that stretches back thousands of years, and they are enjoying the world's most wonderful, versatile, and beautiful material. Vernacular design and old ways of working can still be the best way forward.

> THOSE INVOLVED ARE PART OF A CONTINUOUS TRADITION THAT STRETCHES BACK THOUSANDS OF YEARS, AND THEY ARE ENJOYING THE WORLD'S MOST WONDERFUL, VERSATILE, AND BEAUTIFUL MATERIAL.

When a timber-framed building is erected — as in this photo from a workshop on the island of Gotland, Sweden — tradition requires that a small evergreen tree is attached to the topmost point. The origin of the ancient custom is based in superstition, but I also see it as a symbol of thanks to the tree for all it provides. (Strangely enough, an evergreen is often attached to the last steel beam hoisted up onto a skyscraper in an official "topping out" ceremony.) When the building is roofed, the tree is sawn off flush and the stub remains high inside the peak as a witness.

Acknowledgments

I would like to thank the many clients (or "patrons," as they were referred to historically) whom I have had over the years who allowed me the freedom to be creative with timber and stone on their projects and shared my vision. Thanks go to those clients who graciously opened their homes and barns for photography: Deborah Balmuth and Colin Harrington, Rick and Ginger Scott, Beth and Jeffrey Cohen, Neil and Kathy Chrisman, Allen and Kathleen Williams, David and Helen Lanoue, and Mountain Top Arboretum. I have also had the pleasure of working with some talented individuals during my career in this craft: the late Richard Babcock, Paul Martin, David Carlon, Peter McCurdy, David Bowman, Neil Godden, David Lanoue and his team of expert craftsmen. Though one may work alone, collaboration often produces the finest results.

Lastly, thanks to the good people at Storey, who are a pleasure to work with, and to Hannah Fries, my very competent editor who is no stranger to the craft of timber framing.

Index

Page numbers in *italic* indicate photos or illustrations.

posts, *continued*

oak, fractured, 147, *147*
rot, avoidance of, 89, *89*
tenoned extension for, 47, *47*
waney corner of, 149, *149*
powder post beetles, 51, *51*, 67, *67*
prick marks, 99, *99*
primeval forest. *See* "old-growth" forest
proportion
geometry and, 224
golden, 222–223
proportioning system, 220, *220*
pull boring, 99
purlin plates, 127
purlins, 97
Pythagorean Theorem
3-4-5 triangle and, 219
English barns and, 220–221

R

race knife, 102, *102*
radial-cut wood, 85
rafter foot, 77, *77*
rafters, *17*, 77, *77*, 182–183
rafting of timber, *144*, 145–146, *145*
clues from past, 146, *146*
railroad ties, 61, 63, *63*, 89, 184
railroad trestles, 60
raising techniques, axe marks and, 47, *47*
reading old buildings, 123–124, *124*
recycled materials. *See* reuse of components
red maples, coppiced, 68, *68*
red oak, slip-matching and, 82, *82*
reductions, 106, 182
reuse of components, 23, 30, 32
dendrochronology and, 149
gin poles, 208
legacy of, 34–35, *34, 35*
ripsawing, 140, *140*, 141
riving, 138–139, *138, 139*
rock elm, 126, 127, *127*
roof(s)
board-and-batten, 97
board-on-board, 96–97
European influence, 96–97, *96*
gambrel, 218, *218*
geometry, pitch and, 219, *219*
hammer-beamed, 113

hipped/half-hipped, 96, *96*, 110, *110*, 224, *224*
pitch of, mystery solved, 143
roof purlins, crucks and, 13, *13*
rot, avoidance of, 87–89
round stock, joints in, 77, *77*
rules of thumb, carpentry, 158–161

S

sapwood, 66, 87–89
sawmills, 116, 140–143
circular, 111, *142*, 143
early, 141, *141*
portable, 252, *252*
water-powered, 141, *141*
saws/sawing, 140–143, 189–190
crosscut saws, 190, *190*
by hand, 140, *140*
handsaws, 189–190
ripsawing, 140, *140*, 141, 190, *190*
sawing lines you can see, 189, *189*
seesawing, 140, *140*, 141
scantlings, 114, *114*, 115
scoring, *132*, 133, *133*
Scotch pattern auger, 190, *191*, 193
"scribe rule," 107
scribing, 98–105, *99*
marking, inconsistent, 157, *157*
matching up and, 102, *102*
by the numbers, 99, 103
pin holes, 98–99, *99*
prick marks, 99, *99*
process of, 100–101, *100, 101*
secondary members and, 99
systems of, 98
scribing methods
double cut method, 105, *105*
tumbling method, 104, *104*
sealants, 86
seesawing, 140, *140*, 141
sequencing
clues from past, 152, *152*, 153
math behind, 153–154
sequoias, 53, *53*
shaving horse, 204, *204*
shear-stress concentrations, 182
shed, frame raising for, 44, *44*
shell augers, 190, *191*

Metric Conversion Chart

TO CONVERT	TO	MULTIPLY
inches	centimeters	inches by 2.54
feet	meters	feet by 0.3048
yards	meters	yards by 0.9144
miles	kilometers	miles by 1.609344

Photography Credits *continued from page 4*